Experiments in Green and Sustainable Chemistry
Edited by Herbert W. Roesky and Dietmar K. Kennepohl

Further Reading

Woollins, J. D. (ed.)
Inorganic Experiments
2009
ISBN: 978-3-527-32472-9

Waldmann, H., Janning, P.
Chemical Biology
Learning through Case Studies
2009
ISBN: 978-3-527-32330-2

Schwetlick, K.
Organikum
2009
ISBN: 978-3-527-32292-3

Schwedt, G.
Experimente mit Supermarkt-produkten
Eine chemische Warenkunde
2009
ISBN: 978-3-527-32450-7

Schwedt, G.
Noch mehr Experimente mit Supermarktprodukten
Das Periodensystem als Wegweiser
2009
ISBN: 978-3-527-32476-7

Kerner, N., Lamba, R.
Guided Inquiry Experiments for General Chemistry
Practical Problems and Applications
2007
ISBN: 978-0-471-69842-5

Schwedt, G.
Chemie für alle Jahreszeiten
Einfache Experimente mit pflanzlichen Naturstoffen
2007
ISBN: 978-3-527-31662-5

Macaulay, D. B., Bauer, J. M., Bloomfield, M. M.
General, Organic and Biological Chemistry
An Integrated Approach, Laboratory Experiments
2006
ISBN: 978-0-470-04028-7

Experiments in Green and Sustainable Chemistry

Edited by
Herbert W. Roesky and Dietmar K. Kennepohl

With a Foreword by Jean-Marie Lehn

WILEY-VCH

WILEY-VCH Verlag GmbH & Co. KGaA

The Editors

Prof. Dr. Herbert W. Roesky
Institut für Anorganische Chemie
Georg August Universität
Tammannstr. 4
37077 Göttingen

Dr. Dietmar K. Kennepohl
Department of Chemistry
Athabasca University
University Drive 1
Athabasca, AB TS9 3A3
Kanada

Library of Congress Card No.:
applied for

British Library Cataloguing-in-Publication Data
A catalogue record for this book is available from the British Library.

Bibliographic information published by the Deutsche Nationalbibliothek
The Deutsche Nationalbibliothek lists this publication in the Deutsche Nationalbibliografie; detailed bibliographic data are available on the Internet at http://dnb.d-nb.de.

© 2009 WILEY-VCH Verlag GmbH & Co. KGaA, Weinheim

Typesetting   TypoDesign Hecker GmbH, Leimen
Printing   Strauss GmbH, Mörlenbach
Binding   Litges & Dopf GmbH, Heppenheim
Cover Design   Formgeber, Eppelheim

Printed in the Federal Republic of Germany
Printed on acid-free paper

ISBN: 978-3-527-32546-7

Foreword

Our great appetite for materials and energy is increasing faster than our ability to meet demands. We are seeing limits to the once unlimited resources available to us, and some of the consequences of our current consumption practices are serious. The situation is not trivial, the issues are difficult, and there is no easy way to solve our problems. Yet, sustainable chemistry offers a step in the right direction.

Chemistry has taken on a crucial role in science and society. As the central science, it not only encompasses the simple (such as atomic theory in physics) and the complex (such as life processes in biology), it also is at the heart of many areas that are not necessarily labeled ›chemistry‹. In earth sciences, pharmacy, computing, medicine, materials science, agriculture, nutrition, engineering, and environmental science the practice of chemistry has a profound influence. Because it touches so many of us in everyday life, the science (and art) of transforming matter has become a vital artery both as basic science and in a real tangible sense.

Achieving the transformations of matter with ever-increasing efficiency, selectivity and economy in matter and energy has always occupied the thoughts of chemists, but it is now even more important than ever. Sustainable chemistry builds upon this and covers the whole domain of chemistry. The various experimental examples collected in this book are a wonderful demonstration of a kaleidoscope of chemistry, intertwining the different disciplines.

This contribution by Herbert W. Roesky and Dietmar K. Kennepohl fills a gap. It is particularly welcome for its timeliness. Indeed, it provides a broad coverage of chemistry and highlights the contribution of chemistry to the careful use of the energy and material resources.

The remarkable achievement of this book is its consistent approach to the presentation of green and sustainable chemistry to the younger generation. It can thus be highly recommended to every chemistry teacher.

Strasbourg, July 2008

Jean-Marie Lehn

Contents

Preface

When we want to reach the goal,
Then we want to have the tools.

Immanuel Kant

Edgar F. Smith, an American student, studied chemistry in Göttingen from 1874 to 1876. He did his PhD with Friedrich Wöhler. Later he was president of the University of Pennsylvania for many years and president of the American Chemical Society. He used to relate how, during his time in Wöhler's laboratory, chemical residues were already being recycled:

I had been working in the laboratory for some weeks when I met Professor Wöhler for the first time. During my practical work I had to produce several pounds of phosphoric chloride. For this purpose I had to generate a considerable quantity of other chlorides, and I deposited these residues in the appropriate container. Once when I was emptying my beaker into the garbage container, somebody tapped me on the shoulder. When I looked up I saw that it was Geheimrat Wöhler. He asked me what I was doing. After I had told him all about the chemical preparation, he asked me about the cost of the chemicals I was using. I was not able to calculate them immediately. He then asked me to find this out and look for a chemical merchant who could use the residues, which contained manganese chloride. Soon I discovered that I could sell them provided that they were chemically pure. Wöhler told me to collect the residues separately and to develop a method for getting a chemically pure product. After thinking for a long time I discovered a method of removing the main impurity, iron, and in this way I could earn enough money to buy additional starting materials for my experiments.

(Göttinger Jahresblätter, 1982)

The future is certainly not what we once thought it to be. We live in a world that is becoming increasingly crowded, complex, and rife with new dilemmas. We are certainly concerned for ourselves, but also possess sufficient sense of responsibility to be concerned about the fate of future generations, who will in their turn share these concerns. Today's students are particularly interested in matters that affect their health and the well-being of their planet. So, how do we best equip our chemistry students to deal with the challenges they will meet in the twenty-first century? The concept of Green and Sustainable Chemistry not only offers an excellent opportunity to address some of these concerns, but also provides us with a useful vehicle to advance the way we do chemistry. Since a strong laboratory component is at the heart of many foundation science courses, we present experiments that highlight alternatives and bring out the philosophy behind them. Although this book offers a means to develop laboratory skills and reveal useful techniques, it is ultimately more about encouraging an attitude and approach to chemistry by example and through actual practice.

There is no definitive universal nomenclature for the type of chemistry we demonstrate within this book. We describe it simply as Green and Sustainable Chemistry, others have proposed alternative terms such as Chemistry for the Environment and Environmentally Benign Chemistry. However, Green Chemistry and Sustainable Chemistry seem to be the most widely used and accepted terms. The Organization for Economic Cooperation and Development (OECD) defines Sustainable Chemistry as follows:

Sustainable Chemistry is the design, manufacture, and use of environmentally benign chemical products and processes to prevent pollution, produce less hazardous waste, and reduce environmental and human health risks [1].

One should remember that Sustainable Chemistry is primarily designed for pollution prevention as opposed to waste treatment and control or characterization of chemicals in the environment. The precise terminology will no doubt see further discussion and refinement. However, what is more important is that our collective understanding of the concepts and aims of this chemistry are commonly accepted and will continue to mature.

Some have described the rising popularity of Green and Sustainable Chemistry as some sort of revolution. Indeed, historians may look back at this moment in time and label it as a green revolution much like the agricultural and industrial revolutions that characterized periods in previous centuries. In a revolution there is a sense of wholesale throwing out of the old and replacing it with something new. While much has been achieved in Sustainable Chemistry in a few short years (and there have been specific innovations that could indeed be characterized as revolutionary), we see the process as being both complex and more akin to an evolution. We also note that Green and Sustainable Chemistry is very much integrated within chemistry itself and cannot be treated as a separate discipline. That mainstream nature of Green and Sustainable Chemistry may mean that in years to come it will simply be absorbed into the normal business of what we call chemistry.

So, why is this change occurring? There are several drivers for change towards Green and Sustainable Chemistry that include important considerations like increasing costs, diminishing resource availability, health/environmental legislation, and public perception. The specific drivers and their relative importance are certainly situational, and often their influence is interconnected. Clarke describes and analyzes several key drivers in detail, showing how they stem from three major sources – economic, environmental and social [2]. Chemists as contributors to the economy, as citizens, and as members of society can certainly appreciate the larger challenges and benefits involved as well as anyone else can. However, there is an additional attraction for the synthetic chemist in moving towards Green and Sustainable Chemistry that is very appealing. Noyori [3] expresses this as a pursuit of 'practical elegance' where the core principles of green chemistry are used in developing a synthetic strategy [4]. To the synthetic chemist whose job is essentially to produce valuable substances from baser elements, it is irresistible. The goal of generating the target compound is no longer enough; the manner in which that target compound can be realized is now of great importance. That challenge has awakened innovation and creativity in many chemists and we think it important to convey this passion to students considering chemistry in their studies.

This book has pulled together a diverse feast of teaching experiments from chemists around the globe. The practical real-world experiments illustrate many of the principles of green and sustainable chemistry. Some contributions are from scholars whose primary research is to develop environmentally friendly chemical processes. Others stem from synthetic chemists who have used principles of Green and Sustainable Chemistry in their own disciplinary research. All are chemical educators who want to share their knowledge and passion for Green and Sustainable Chemistry. We are very thankful for their contributions.

Herbert W. Roesky
Dietmar K. Kennepohl

Caution!

The authors of these experiments have taken great care to describe the nature of the substances and equipment employed. As with all chemistry experiments care should be taken in handling hazardous materials in an appropriate manner. The experiments are intended for students working in a safe laboratory environment under qualified supervision.

References

1 *Proceedings of the OECD Workshop on Sustainable Chemistry, Part 3*, Venice, October 15–17, 1998; OECD: **1999**.
http://www.olis.oecd.org/olis/1999doc.nsf/LinkTo/NT00000FDA/$FILE/09E98688.PDF (accessed 2/19/08).

2 J.H. Clark, *Green Chem.* **2006**, 8, 17.
3 R. Noyori, *Chem. Commun.* **2005**, 1807.
4 P. Anastas, J. Warner *Green Chemistry: Theory and Practice;* Oxford University Press: New York, **1998**.

List of Contributors

Mohammed Abid
Department of Chemistry
University of Massachusetts Boston
100 Morrissey Blvd.
Boston, MA 02125
USA

Lutz Ackermann
Georg August Universität Göttingen
Institut für Organische und
Biomolekulare Chemie
Tammannstrasse 2
37077 Göttingen
Germany

Angelo Albini
Department of Organic Chemistry
University of Pavia
Via Taramelli 10
27100 Pavia
Italy

Marius Andruh
Inorganic Chemistry Laboratory
Faculty of Chemistry
University of Bucharest
Str. Dumbrava Rosie nr. 23
020464-Bucharest
Romania

Didier Astruc
Institut des Sciences Moléculaires
UMR CNRS N°5255
Université Bordeaux I
351 Cours de la Libération
33405 Talence Cedex
France

Hans-Dieter Barke
Institut für Didaktik der Chemie
Westfälische Wilhelms Universität
Münster
Fliednerstraße 21
48149 Münster
Germany

Andreas Bösmann
Institut für Chemische Reaktions-
technik
Universität Erlangen-Nürnberg,
Egerlandstrasse 3
91058 Erlangen
Germany

C. Christian Brazel
Georg August Universität Göttingen
Institut für Organische und
Biomolekulare Chemie
Tammannstrasse 2
37077 Göttingen
Germany

John P. Canal
Department of Chemistry
Simon Fraser University
8888 University Drive
Burnaby, BC V5A 1S6
Canada

Mónica Carril
Kimika Organikoa II Saila
Zientzia eta Teknologia Fakultatea
Euskal Herriko Unibertsitatea
PO Box 644
48080 Bilbao
Spain

Ji Chen
Changchun Institute of Applied
Chemistry
Chinese Academy of Sciences
Changchun 130022
China

Claus H. Christensen
Center for Sustainable and Green
Chemistry
Department of Chemistry
Building 206
Technical University of Denmark
2800 Lyngby
Denmark

James H. Clark
Green Chemistry Center of
Excellence
The University of York
Heslington, York YO10 5DD
UK

Jason A. C. Clyburne
Department of Chemistry
Saint Mary's University
923 Robie Street
Halifax, NS B3H 3C3
Canada

Alan H. Cowley
Department of Chemistry &
Biochemistry
University of Texas at Austin
Austin, Texas 78712
USA

Emma Coyle
School of Chemical Sciences and
NCSR
Dublin City University
Dublin 9
Ireland

Michael P. DeVore
Department of Chemistry
Laboratory for Environmentally
Friendly Organic Synthesis
Illinois Wesleyan University
Bloomington, IL 61701
USA

Rodrigue Djeda
Institut des Sciences Moléculaires
UMR CNRS N°5255
Université Bordeaux I
351 Cours de la Libération
33405 Talence Cedex
France

Adinarayana Doddi
Department of Chemistry,
Indian Institute of Technology
Madras, Chennai – 600 036
India

Esther Domínguez
Kimika Organikoa II Saila
Zientzia eta Teknologia Fakultatea
Euskal Herriko Unibertsitatea
PO Box 644
48080 Bilbao
Spain

Daniele Dondi
Department of General Chemistry
University of Pavia
Via Taramelli 12
27100 Pavia
Italy

Gordon W. Driver
Department of Chemistry and
Biochemistry
University of Regina
3737 Wascana Parkway
Regina, SK S4S 0A2
Canada

Bobby D. Ellis
Department of Chemistry
University of California – Davis
Davis, CA 95616
USA

Maurizio Fagnoni
Department of Organic Chemistry
University of Pavia
Via Taramelli 10
27100 Pavia
Italy

Megumi Fujita
Department of Chemistry
University of West Georgia
Carrollton, GA 30118
USA

Anthony Fusco
Department of Chemistry
Northeastern University
Boston, MA 02115
USA

Ruxandra Gheorghe
Inorganic Chemistry Laboratory
Faculty of Chemistry
University of Bucharest
Str. Dumbrava Rosie nr. 23
020464-Bucharest
Romania

Yury Gorbanov
Center for Sustainable and Green
Chemistry
Department of Chemistry
Building 206
Technical University of Denmark
2800 Lyngby
Denmark

Robert A. Gossage
Department of Chemistry & Biology:
Faculty of Engineering, Architecture
& Science
Ryerson University
350 Victoria Street
Toronto, ON M5B 2K3
Canada

Hansjörg Grützmacher
ETH Zürich
Department of Chemistry and
Applied Biosciences
Wolfgang Pauli Strasse 10
HCI H 131
8093 Zürich
Switzerland

Jack M. Harrowfield
Institut de Science et d'Ingénierie
Supramoléculaires
Université Louis Pasteur, 8 allée
Gaspard Monge
Strasbourg 67083
France

Markus Hölscher
Institut für Technische und Makro-
molekulare Chemie
RWTH Aachen University
Worringerweg 1
52074 Aachen
Germany

Matthew G. Huddle
Department of Chemistry
Laboratory for Environmentally
Friendly Organic Synthesis
Illinois Wesleyan University
Bloomington, IL 61701
USA

Jonathan G. Huddleston
Millipore Bioprocessing Ltd.
Medomsley Road
Consett, County Durham DH8 6SZ
UK

Keith E. Johnson
Department of Chemistry and
Biochemistry
University of Regina
3737 Wascana Parkway
Regina, SK S4S 0A2
Canada

Kieran Joyce
School of Chemical Sciences and NCSR
Dublin City University
Dublin 9
Ireland

Michael J. Katz
Department of Chemistry
Simon Fraser University
8888 University Drive
Burnaby, BC V5A 1S6
Canada

Dietmar K. Kennepohl
Athabasca University
1 University Drive
Athabasca, AB T9S 3A3
Canada

Horst Kisch
Institute of Inorganic Chemistry
Department of Chemistry and
Pharmacy
University of Erlangen-Nürnberg

Søren K. Klitgaard
Center for Sustainable and Green
Chemistry
Department of Chemistry
Building 206
Technical University of Denmark
2800 Lyngby
Denmark

George A. Koutsantonis
Chemistry, M313, School of
Biomedical
Biomolecular and Chemical Sciences
The University of Western Australia
35 Stirling Highway
Crawley, 6009, WA
Australia

Ingo Krossing
Institut für Anorganische und
Analytische Chemie
Albert Ludwigs Universität Freiburg
Albertstrasse 21
79104 Freiburg i. Br.
Germany

Shainaz M. Landge
Department of Chemistry
University of Massachusetts Boston
100 Morrissey Blvd.
Boston, MA 02125
USA

Nicholas E. Leadbeater
Department of Chemistry
University of Connecticut
55 North Eagleville Road
Storrs, CT 06269-3060
USA

Julie Lefebvre
Department of Chemistry
Simon Fraser University
8888 University Drive
Burnaby, BC V5A 1S6
Canada

Walter Leitner
Institut für Technische und
Makromolekulare Chemie
RWTH Aachen University
Worringerweg 1
52074 Aachen
Germany

Daniel B. Leznoff
Department of Chemistry
Simon Fraser University
8888 University Drive
Burnaby, BC V5A 1S6
Canada

Rafael Luque
Green Chemistry Center of
Excellence
The University of York
Heslington, York YO10 5DD
UK

Alexander Lygin
Georg August Universität Göttingen
Institut für Organische und
Biomolekulare Chemie
Tammanstrasse 2
37077 Göttingen
Germany

Patricia Ann Mabrouk
Department of Chemistry
Northeastern University
Boston, MA 02115
USA

Charles L. B. Macdonald
Department of Chemistry and
Biochemistry,
University of Windsor
401 Sunset Ave
Windsor, ON N9B 3P4
Canada

Duncan J. Macquarrie
Green Chemistry Center of Excel-
lence
The University of York
Heslington, York YO10 5DD
UK

Augustin M. Madalan
Inorganic Chemistry Laboratory
Faculty of Chemistry
University of Bucharest
Str. Dumbrava Rosie nr. 23
020464-Bucharest
Romania

Jochen Mattay
Organische Chemie I
Fakultät für Chemie
Universität Bielefeld
Postfach 10 01 31
33501 Bielefeld
Germany

Cynthia McGowan
Department of Chemistry
University of Connecticut
55 North Eagleville Road
Storrs, CT 06269-3060
USA

Armin de Meijere
Georg August Universität Göttingen
Institut für Organische und
Biomolekulare Chemie
Tammanstrasse 2
37077 Göttingen
Germany

Franc Meyer
Georg August Universität Göttingen
Institut für Anorganische Chemie
Tammannstrasse 4
37077 Göttingen
Germany

Ram S. Mohan
Department of Chemistry
Laboratory for Environmentally
Friendly Organic Synthesis
Illinois Wesleyan University
Bloomington, IL 61701
USA

Michael Oelgemöller
School of Chemical Sciences and NCSR
Dublin City University
Dublin 9
Ireland

Catia Ornelas
Institut des Sciences Moléculaires
UMR CNRS N° 5255
Université Bordeaux I
351 Cours de la Libération
33405 Talence Cedex
France

Jiri Pinkas
Department of Chemistry
Masaryk University
Kotlarska 2
61137 Brno
Czech Republic

Stefano Protti
Department of Organic Chemistry
University of Pavia
Via Taramelli 10
27100 Pavia
Italy

Ines Raabe
Institut für Anorganische und
Analytische Chemie
Albert Ludwigs Universität Freiburg
Albertstrasse 21
79104 Freiburg i. Br.
Germany

Taramatee Ramnial
Department of Chemistry
Simon Fraser University
8888 University Drive, Burnaby
BC V5A 1S6
Canada

M.N. Sudheendra Rao
Department of Chemistry
Indian Institute of Technology
Madras, Chennai – 600 036
India

Gregor Reeske
Department of Chemistry &
Biochemistry
University of Texas at Austin
Austin, Texas 78712
USA

Andreas Reisinger
Institut für Anorganische und
Analytische Chemie
Albert Ludwigs Universität Freiburg
Albertstrasse 21
79104 Freiburg i. Br.
Germany

Anders Riisager
Center for Sustainable and Green
Chemistry
Department of Chemistry
Technical University of Denmark
Kemitorvet, Building 207
2800 Kgs. Lyngby
Denmark

Herbert W. Roesky
Georg August Universität Göttingen
Institut für Anorganische Chemie
Tammannstrasse 4
37077 Göttingen
Germany

Peter W. Roesky
bFreie Universität Berlin
Institut für Chemie und Biochemie
Fabeckstrasse 34-36
14195 Berlin
Germany

Robin D. Rogers
Center for Green Manufacturing and
Department of Chemistry
Alabama Institute for Manufacturing
Excellence
The University of Alabama
Tuscaloosa, AL 35487
USA

Jaime Ruiz
Institut des Sciences Moléculaires
UMR CNRS N°5255
Université Bordeaux I
351 Cours de la Libération
33405 Talence Cedex
France

Raul SanMartin
Kimika Organikoa II Saila
Zientzia eta Teknologia Fakultatea
Euskal Herriko Unibertsitatea
PO Box 644
48080 Bilbao
Spain

Hartmut Schönberg
ETH Zürich
Department of Chemistry and
Applied Biosciences
Wolfgang Pauli Strasse 10
HCI H 131
8093 Zürich
Switzerland

Dong-Kyun Seo
Department of Chemistry and
Biochemistry
Arizona State University
Tempe, AZ 85287-1604
USA

Tameka L. Shamery
Center for Green Manufacturing and
Department of Chemistry
Alabama Institute for Manufacturing
Excellence
The University of Alabama
Tuscaloosa, AL 35487
USA

Scott K. Spear
Alabama Institute for Manufacturing
Excellence
The University of Alabama
Tuscaloosa, AL 35487
USA

Dirk A. Spiegl
Georg August Universität Göttingen
Institut für Organische und
Biomolekulare Chemie
Tammannstrasse 2
37077 Göttingen
Germany

Michael Stollenz
Georg August Universität Göttingen
Institut für Anorganische Chemie
Tammannstrasse 4
37077 Göttingen
Germany

Esben Taarning
Center for Sustainable and Green
Chemistry
Department of Chemistry
Building 206
Technical University of Denmark
2800 Lyngby
Denmark

Lutz F. Tietze
Georg August Universität Göttingen
Institut für Organische und
Biomolekulare Chemie
Tammannstrasse 2
37077 Göttingen
Germany

Béla Török
Department of Chemistry
University of Massachusetts Boston
100 Morrissey Blvd.
Boston, MA 02125
USA

Rubén Vicente
Georg August Universität Göttingen
Institut für Organische und
Biomolekulare Chemie
Tammannstrasse 2
37077 Göttingen
Germany

Sandra Walendy
Institut für Technische und Makro-
molekulare Chemie
RWTH Aachen University
Worringerweg 1
52074 Aachen
Germany

Stefan Zimmermann
Universität Karlsruhe (TH)
Institut für Anorganische Chemie
Engesserstrasse 15
76128 Karlsruhe
Germany

Part I
Catalysts

1

Clean Friedel–Crafts Acylation of Anisole under Microwave Irradiation

Rafael Luque, James H. Clark and Duncan J. Macquarrie

A Friedel–Crafts acylation involves the electrophilic substitution of a hydrogen atom on an aromatic ring with a R–CO– group, forming an aromatic ketone. This reaction is a swift route to important intermediates and high added-value chemicals that find applications in the fine chemical, fragrances and pharmaceutical industries. Traditionally, the acylation has been carried out in a homogeneous phase using acyl halides or anhydrides as acetylating agents and Lewis acids, including $AlCl_3$, $FeCl_3$ and $ZnCl_2$, as catalysts. The inherent drawbacks of all homogeneous systems are the production of significant quantities of hazardous waste (e.g. salts, acids, etc.), necessitating thorough washing and neutralization, and the difficulties in the catalyst recovery. The use of water to quench the reaction leads to large amounts of gaseous and aqueous waste as well as loss of catalyst. The process has therefore a large E factor, as all the $AlCl_3$ employed is wasted. A combination of green technologies is therefore needed to improve the green credentials of the reaction. Heterogeneous catalysis offers a more environmentally benign approach, overcoming problems with catalyst recycling and generation of waste. Novel reaction technologies, including alternative reaction media (e.g. solventless environments, ionic liquids, supercritical CO_2) and microwaves also enhance reaction rates and reduce the energy consumption, providing better yields and/or selectivities in favor of the desired products. The aim of this practical is to carry out a clean synthesis of 4-methoxyacetophenone from anisole (Scheme 1) using a solid acid catalyst under microwave irradiation.

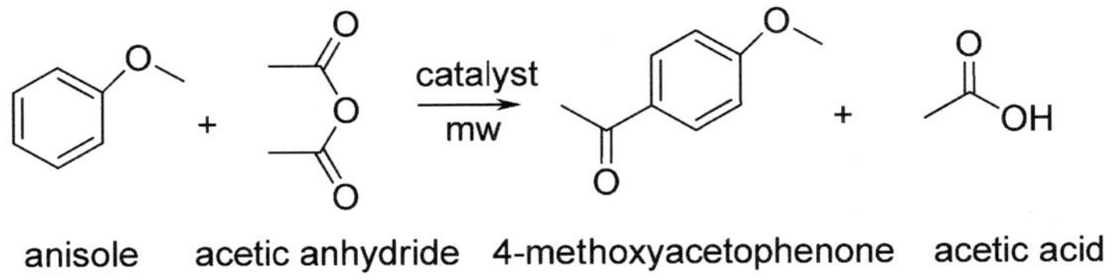

| anisole | acetic anhydride | 4-methoxyacetophenone | acetic acid |

Scheme 1 Friedel–Crafts acylation of anisole with acetic anhydride under microwave irradiation.

Preparation of 4-methoxyacetophenone

Apparatus CEM microwave apparatus, microwave tube, magnetic stirrer, Varian gas chromatograph fitted with a 30 m DB-5 column (0.25 mm internal diameter), furnace, safety glasses, laboratory coat, protective gloves.

Chemicals Anisole (20 mmol, 2.16 g); acetic anhydride (30 mmol, 3.06 g); a choice of a solid acid catalyst between beta-zeolite/Nafion®/sulfated zirconia. The solid acid catalyst (depending on availability) should be activated prior to its use as the catalyst in the reaction, with the exception of the Nafion®. Beta-zeolite and sulfated zirconia (Süd-Chemie) are activated in a furnace at 550 °C for 3 h. Nafion® (Dupont) is used directly as purchased.

Attention! Safety glasses and protective gloves must be worn at all times when handling chemicals.

Caution! Because of their toxicity and volatility, care should be taken to avoid inhalation of acetic anhydride and anisole. Avoid skin contact with the chemicals, including the solid acid catalysts. The preparation of the microwave mixture should be carried out inside the fume hood.

Procedure An oven-dried microwave tube is charged with 20 mmol anisole (2.16 g), 30 mmol acetic anhydride (3.06 g) and 0.6 g solid acid. The mixture is then placed in the microwave reactor and irradiated at 200 W for 5 min until the reaction reaches a temperature of 150 °C. The final mixture after the reaction is then filtered off to remove the solid acid catalyst and a liquid sample is taken for gas chromatography (GC) analysis. The analysis areas under the different GC peaks are used to work out the experimental anisole:products ratio and are compared to the expected anisole:products ratio in order to calculate the molar amount of product generated and hence the conversion of the starting material. The products obtained are ortho- (retention time 7.8 min) and para- (retention time 8.4 min) isomers and minor quantities of polyacetylated anisoles including diacetylated (retention time 10.7 min) and triacetylated (retention time 11.6 min). Selectivity to the desired product (i.e. 4-methoxyacetophenone) is also calculated from GC data and included in Table 1.

Waste Disposal Solid catalysts can be recovered and washed to be reused in the reaction. Chemicals should be disposed of as either hydrocarbon waste (catalytic reactions) or water miscibles (preparation of the aminopropylated silica).

Explanation

In general, the microwave reaction produced high yields of acetylated products using sulfated zirconia or beta zeolite as catalysts (Table 1). Quantitative conversion of the starting material was found after 5 min of microwave irradiation with a very high selectivity to the para isomer. The zeolite material was found to be the most active and selective material in the acetylation reaction, being reusable after the reaction without a great loss in activity (Table 1). This novel technology has been shown to cut reaction times down by orders of magnitude (5 min compared to 5–24 h reaction in the conventional heating protocol), improving the energy efficiency of the reaction.

Table 1 Activity of different solid acid catalysts in the Friedel–Crafts acylation of anisole. (Reaction conditions: 20 mmol anisole, 30 mmol acetic anhydride, 0.6 g catalyst, 200 W, 5 min, 150 °C; [a]40 mmol anisole, 20 mmol acetic anhydride, 45 mmol AlCl$_3$, 200 W, 30 min, 90 °C.)

Catalyst	Conversion (mol%)	Selectivity para- (mol%)
Blank	–	–
Nafion®	30	>95
Sulfated zirconia	85	>99
Beta zeolite	>95	>99
Beta zeolite (reused)	>95	>99
AlCl$_3$[a]	80	>95

The anisole conversions achieved with the solid acid catalyst exceeded those of $AlCl_3$, and all the major drawbacks of the homogeneous protocol were observed, including the generation of 60 g waste created for every gram of product produced.

In summary, the acetylation of anisole can be performed in a more efficient and greener way by using a microwave approach employing solid acid catalysts replacing the traditional $AlCl_3$ Lewis acid. The microwave protocol significantly reduced reaction times and energy usage, and improved yields and selectivities, making the reaction simple and conveniently carried out.

References

1 M. Hino and K. Arata, *Chem Commun.* **1995**, 112.
2 Y. Ma, Q. Wang, W. Jiang, B. Zuo, *Appl. Catal. A* **1997**, *165*, 199.
3 U. Freese, F. Heinrich, F. Roessner, *Catal. Today* **1999**, *49*, 237.
4 B. Bachiller-Baeza, J.A. Anderson, *J. Catal.* **2004**, *228*, 225.
5 C. Hardacre, S.P. Katdare, D. Milroy, P. Nancarrow, D.W. Rooney, J.M. Thompson, *J. Catal.* **2004**, *227*, 44.
6 M.L. Kantman, K.V.S. Ranganath, M. Sateesh, K.B.S. Kumar, B. Choudary, *J. Mol. Catal. A* **2005**, *255*, 15.
7 V. Budarin, J.H. Clark, R. Luque, D.J. Macquarrie, *Chem. Commun.* **2007**, 534.

2

Knoevenagel Condensation Using Aminopropylated Silicas

Rafael Luque, James H. Clark and Duncan J. Macquarrie

The development of solid bases is of extreme interest, as these can replace the liquid bases that are currently employed in various industrial processes. Thus, the organic reactions catalyzed by solid bases are a relatively new and interesting area in organic synthesis. Of relevant interest are those focused on the preparation of fine chemicals. The Knoevenagel condensation is one of the most useful C–C bond-forming reactions for the production of fine chemicals. It is a cross-aldol condensation between an aldehyde or a ketone and a methylene activated substrate to yield an α,β-unsaturated carboxylic ester. Several heterogeneous base catalysts have been reported to be active in the reaction including oxides, zeolites, and mesoporous materials. Mesoporous silicas, featuring a unique pore distribution, ease of tuneability and availability, appear to be one of the best candidates as support material and/or catalyst for this reaction. We have reported the preparation of aminopropylated silicas for a wide range of base-catalyzed reactions. The aim of this practical is to carry out the preparation of various α,β-unsaturated compounds by the Knoevenagel condensation of various ketones with ethyl cyanoacetate (Scheme 1) using aminopropylated silica as heterogeneous base catalyst. Aldehydes and ketones both smoothly react with ethyl cyanoacetate, chosen as a carbon acid of moderate activity and synthetic utility (Scheme 1).

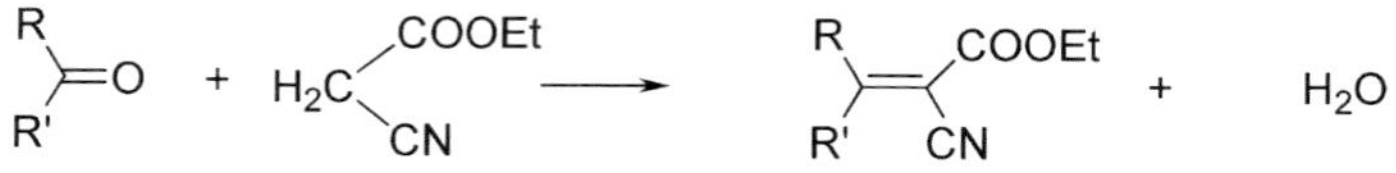

| ketone | ethyl cyanoacetate | α,β-unsaturated carboxylic ester |

Scheme 1 Knoevenagel condensation of various ketones with ethyl cyanoacetate catalyzed by aminopropylated silicas.

Preparation of Knoevenagel Condensation Products

Apparatus

Hot plate, round-bottomed flask, magnetic stirrer, Dean and Stark trap, gas chromatograph fitted with a DB-5 column (0.25 mm internal diameter), furnace, safety glasses, laboratory coat, protective gloves.

Chemicals

A range of ketones (20 mmol, depending on availability), ethyl cyanoacetate (20 mmol, 2.13 mL), toluene or cyclohexane as solvents and aminopropylated silica (APS) as solid base catalyst are needed. The solid base catalyst can be prepared as follows. Tetraethoxysilane (TEOS, 18.75 g, 0.09 mol) and triethoxy(3-aminopropyl)silane (AMPS, 1.79 g, 0.01 mol) are added, separately, to a stirred mixture of ethanol (41 g), distilled water (53 g) and n-dodecylamine (5.09 g) at room temperature. After 10–15 min the turbid solution becomes milky, and stirring must continue for at least 3 h, yielding a thick white suspension. This suspension is then filtered and n-dodecylamine can be removed by heating the solid (ca. 15 g) at reflux in absolute ethanol (100 mL) for 3 h. This extraction has to be repeated three times. After the final filtration, the product is dried at 120 °C and atmospheric pressure to yield a fine white solid. The typical loading of aminopropyl groups is 1–2 mmol g^{-1}.

Attention!

Safety glasses and protective gloves must be worn at all times when handling chemicals.

Caution!

Because of their toxicity, care should be taken to avoid inhalation or direct skin contact with the reagents (ketones, ethyl cyanoacetate, toluene and cyclohexane) and the solid base catalyst. The preparation of the reaction mixture should be carried out inside the fumehood.

Procedure

An oven-dried 50 mL round-bottomed flask is charged with 20 mmol ketone, 20 mmol ethyl cyanoacetate (2.13 mL), 0.25 g APS and 25 mL toluene/cyclohexane. n-Dodecane (1.2 mmol, 0.2 g) is used as internal standard. The mixture is then heated at solvent reflux (110 °C for toluene and 82 °C for cyclohexane, respectively) for 0.5 to 4 h. A Dean and Stark apparatus can be used to remove water from the reaction; this may slightly speed up the reaction. The final mixture after reaction is filtered off to remove the solid base catalyst and a liquid sample is taken for gas chromatography (GC) analysis. The areas under the different GC peaks are used to work out the experimental reagent:product ratio, and this is compared to the expected reagent:product ratio in order to calculate the molar amount of product generated and hence the conversion of the starting material. The only product found in the reaction (α,β-unsaturated carboxylic ester) is due to the Knoevenagel condensation. The results are included in Table 1.

Waste Disposal Solid catalysts can be recovered and washed to be reused in the reaction. Chemicals should be disposed as either hydrocarbon waste (catalytic reactions) or water miscibles (preparation of the aminopropylated silica).

Explanation

In general, the Knoevenagel condensation of the various ketones with ethyl cyanoacetate afforded quantitative yields of the main condensation product in a few hours of reaction (Table 1). The aminopropylated silica is a very active and selective catalyst for the reaction, even at relatively low loading of aminopropyl groups. The applicability of the amines as catalysts is related to the reaction

Table 1 Activity of the aminopropylated silica in the Knoevenagel condensation of various ketones with ethyl cyanoacetate. (Reaction conditions: 20 mmol aldehyde/ketone, 20 mmol ethyl cyanoacetate (20 mmol, 2.13 mL), 0.25 g APS, 25 mL solvent, 1.2 mmol n-dodecane.)

Substrate	Solvent	Temp. (°C)	Time (h)	Yield (%)
	toluene	110	2	>90
	toluene	110	4	>95
	cyclohexane	82	2	>95
	cyclohexane	82	1	>90

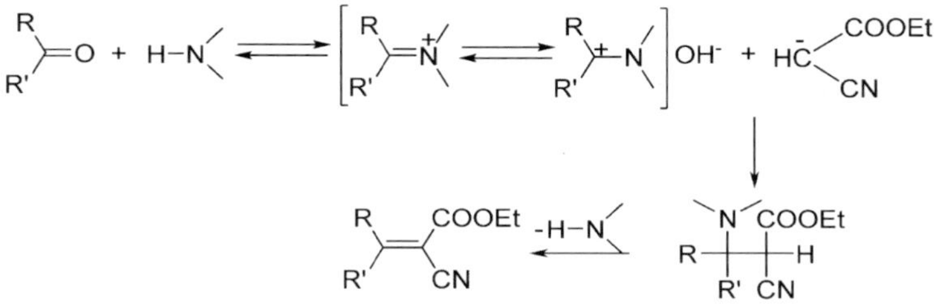

Scheme 2 Reaction mechanism of the Knoevenagel condensation of ketones and ethyl cyanoacetate using amine groups as catalysts.

mechanism, which is believed to involve the formation of iminium intermediate species (Scheme 2) when amine groups are employed as catalysts. This is referred to as the Hann-Lapworth mechanism.

In summary, the Knoevenagel condensation of various ketones with ethyl cyanoacetate can be easily performed employing a solid base catalyst. The reported protocol afforded the quantitative and selective formation of different α,β-unsaturated carboxylic esters (important intermediates in the preparation of fine chemicals) in a few hours.

References

1 E. Knoevenagel, *Ber. Dtsch. Chem. Ges.* **1894**, *27*, 2345.

2 G. Jones, *Org. React.* **1967**, *15*, 204.

3 D.J. Macquarrie, *Chem. Commun.* **1996**, 1961.

4 D.J. Macquarrie, D. B. Jackson, *Chem. Commun.* **1997**, 1781.

5 K.A. Utting, D. J. Macquarrie, *New J. Chem.* **2000**, *24*, 591.

6 D.J. Macquarrie, *Green Chem.* **1999**, *1*, 195.

7 D.J. Macquarrie, D.B. Jackson, S. Tailland, K.A. Utting, *J. Mater. Chem.* **2001**, *11*, 1843.

8 T. Jackson, J.H. Clark, D.J. Macquarrie, J.H. Brophy, *Green Chem.* **2004**, 6, 193.

3

Copper-Catalyzed Arylation of Thiols in Water

Mónica Carril, Raul SanMartin and Esther Domínguez

Because of the presence of diaryl sulfides in a great number of pharmaceuticals, biologically active molecules and polymeric materials, much effort has been devoted to their synthesis [1]. In general, copper catalysts provide an economically and often environmentally advantageous alternative to other methodologies designed to form C_{aryl}–O and C_{aryl}–N bonds [2], but the extension of copper catalysis to the formation of the C_{aryl}–S linkage is still a field scarcely explored [2, 3]. Moreover, nonrecyclable catalytic systems, potentially harmful organic solvents and strong bases are common features of most of the reported methods [3]. In recent years, water has become the solvent of choice for many organic transformations, since in addition to its well-known environmental benignity, aspects like nontoxicity, low cost, availability and greater chemoselectivity compared to organic solvents are considered [4].

Following our research on sustainable chemical processes, we reported the synthesis of a series of benzofurans by an intramolecular copper-catalyzed O-arylation of ketones performed in water [5]. Encouraged by this successful application of »on-water chemistry« [4], we extended this protocol (based on the use of a catalytic amount of a copper salt, a 1,2-diamine that plays the role of ligand and base, and water as the only solvent) to the formation of C_{aryl}–S bonds by coupling between thiophenols and aryl halides, thus providing a novel entry to diaryl sulfides. Moreover, the aqueous layer containing the catalyst could be reused up to 2 times without losing its catalytic activity [6].

Two examples of copper-catalyzed arylation of thiols

Preparation of 4-chlorophenyl phenyl sulfide

$$\text{PhSH} + \text{4-chloroiodobenzene} \xrightarrow[\substack{\text{CuCl (8.5 mol\%)} \\ H_2O,\ 120\ ^\circ C,\ 8\ h}]{\substack{\text{trans-1,2-diaminocyclohexane}}} \text{4-chlorophenyl phenyl sulfide (97\%)}$$

Apparatus	10 mL Schlenk flask with septum, magnetic stirrer, Ar inlet (balloon), syringes (1 mL), stainless steel needles, standard equipment for gravity column chromatography, rotary evaporator, safety glasses, laboratory coat, protective gloves.
Chemicals	CuCl, 4-chloroiodobenzene, PhSH, *trans*-1,2-diaminocyclohexane, H_2O, dichloromethane, hexane, Na_2SO_4.
Attention!	Safety glasses and protective gloves must be worn at all times.
Caution!	Because of their moderate toxicity and volatility, care should be taken to avoid inhalation of thiophenol and *trans*-1,2-diaminocyclohexane, or contact of these chemicals and copper(I) chloride or 4-chloroiodobenzene with the skin. All reactions should be carried out in a well-ventilated hood behind a protective shield. Pressure builds up when the Schlenk flask is heated.
Procedure	A Schlenk flask equipped with a magnetic stirrer, a septum, and an argon balloon is charged with CuCl (4.0 mg, 0.040 mmol), 4-chloroiodobenzene (227.2 mg, 0.94 mmol), *trans*-1,2-diaminocyclohexane (0.21 mL, 1.83 mmol), water (6.1 mL) and thiophenol (0.05 mL, 0.47 mmol). Then the flask is sealed under a positive pressure of argon and the obtained violet solution is heated overnight at 120 °C. The aqueous layer is extracted with dichloromethane, dried over anhydrous sodium sulfate and concentrated *in vacuo*. The residue is then purified by flash chromatography (10% CH_2Cl_2/hexane) to give 4-chlorophenyl phenyl sulfide (102.7 mg, 99%) [7] as a colorless liquid.

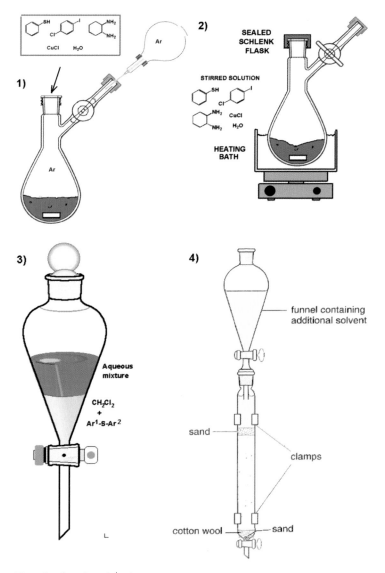

Figure 1 Experimental set-ups.

Recycling of the aqueous solution containing the catalyst: After extraction with CH_2Cl_2, the recovered aqueous layer is placed in a Schlenk flask under argon, and 4-chloroiodobenzene (227.2 mg, 0.94 mmol), *trans*-diaminocyclohexane (0.17 mL, 1.42 mmol) and thiophenol (0.05 mL, 0.47 mmol) are added. The obtained mixture is heated at

120 °C overnight and the isolation of the product is accomplished as described previously. This procedure, employing identical amounts for each reactant, is repeated three times, obtaining 4-chlorophenyl phenyl sulfide with 99%, 95% and 97% yields respectively.

Characterization data: ^{1}H NMR (300 MHz, CDCl$_3$): δ 7.37–7.22 (m, 9H) ppm. ^{13}C NMR (75 MHz, CDCl$_3$): δ 135.1, 134.6, 132.9, 131.9, 131.3, 129.3, 129.2, 127.4 ppm. HRMS(EI) *m/z:* Found: 220.0116. C$_{12}$H$_9$SCl requires 220.0113. LRMS(EI) *m/z:* 222 (M^{2+}, 7%), 220 (M, 40%), 185 (100%).

Preparation of 2-aminophenyl phenyl sulfide

Apparatus	10 mL Schlenk flask with septum, magnetic stirrer, Ar inlet (balloon), syringes (1 mL), stainless steel needles, standard equipment for gravity column chromatography, rotary evaporator, safety glasses, laboratory coat, protective gloves.
Chemicals	CuCl, NaI, 4-bromoaniline, PhSH, *trans*-1,2-diaminocyclohexane, H$_2$O, silica gel 60 (particle size 230–400 mesh).
Attention!	Safety glasses and protective gloves must be worn at all times.
Caution!	Because of their moderate toxicity and volatility, care should be taken to avoid inhalation of thiophenol and *trans*-1,2-diaminocyclohexane, or contact of these chemicals and copper(I) chloride or 4-bromoaniline with the skin. All reactions should be carried out in a well-ventilated hood behind a protective shield. Pressure builds up when the Schlenk flask is heated.
Procedure	*Br/I exchange and subsequent S-arylation:* A Schlenk flask equipped with a magnetic stirrer, a septum, and an argon balloon is charged with CuCl (4.0 mg, 0.04 mmol), 4-bromoaniline (167.3 mg, 0.94 mmol), *trans*-1,2-diaminocyclohexane (20 μL, 0.17 mmol), NaI

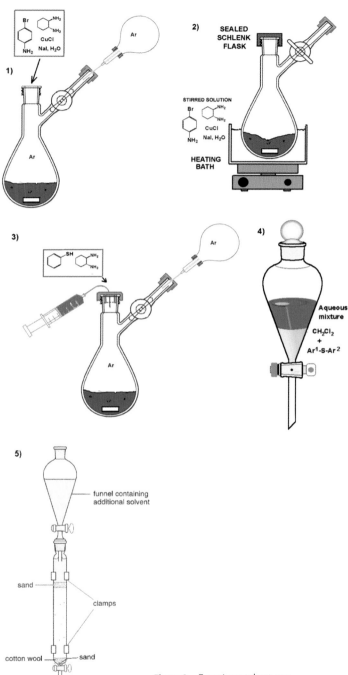

Figure 2 Experimental set-ups.

(284.2 mg, 1.89 mmol) and water (6.3 mL). Then, the flask is sealed under a positive pressure of argon and the obtained violet solution is heated at 120 °C. After 24 h, the mixture is allowed to reach room temperature and then *trans*-1,2-diaminocyclohexane (0.20 mL, 1.67mmol) and thiophenol (0.05 mL, 0.47 mmol) are added under argon, the flask is sealed again and the solution is heated overnight at 120 °C. After cooling, the aqueous layer is extracted with dichloromethane, dried over anhydrous sodium sulfate and concentrated *in vacuo*. The residue is then purified by flash chromatography (10% CH_2Cl_2/hexane) to give 2-aminophenyl phenyl sulfide (79.6 mg, 84%) [8] as a white powder.

Characterization data: ^{1}H NMR (300 MHz, $CDCl_3$): δ 7.33 (d, $J = 8.44$ Hz, 2H), 7.26–7.09 (m, 5H), 6.68 (d, $J = 8.46$ Hz, 2H), 3.81 (bs, 2H) ppm. ^{13}C NMR (75 MHz, $CDCl_3$): δ 146.7, 139.4, 135.9, 128.7, 127.2, 125.2, 120.3, 115.8 ppm. HRMS(EI) *m/z:* Found: 201.0611. $C_{12}H_{11}NS$ requires 201.0612. LRMS(EI) *m/z:* 201 (M, 100%).

Waste Disposal The basic aqueous solution containing the copper complex and traces of starting thiolate should be disposed of in the aqueous waste container. The contaminated filter paper should be placed in the solid waste container. Contaminated silica gel and sodium sulfate must be disposed of as hazardous waste in a specially designed, sealable bag or container.

Explanation

A range of different 1,2-diamines have been assayed in order to optimize this aqueous *S*-arylation protocol, and this showed that *trans*-1,2-diaminocyclohexane (3.9 equivalents) is the most convenient ligand/base in combination with CuCl (8.5 mol%). The presented copper-catalyzed *S*-arylation methodology can be applied not only to aryl iodides but also to bromides bearing electron-withdrawing substituents. However, in order to perform the target transformation with an aryl bromide bearing electron-donating groups like 4-amino (see the latter example), an *in situ* halogen exchange [9] performed in water has been devised.

Most of the published work on catalyst reutilization relies on the separation of the metal complex from the reaction medium, and often reactivation is also re-

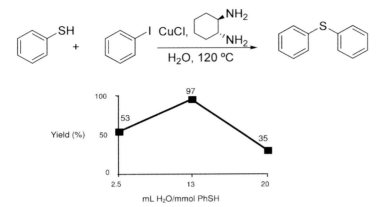

Yield (%)

100
97
53
50
35
0

2.5   13   20

mL H₂O/mmol PhSH

Figure 3 Optimal amount of water.

quired [10]. However, here not only is the copper complex recycled, but so also is the aqueous reaction medium itself (containing the active catalyst).

Finally, regarding the reaction mechanism, it is surprising to find out that, unlike other »on-water« processes, the relative amount of water is crucial for the success of the reaction. Indeed, after different dilution trials in the coupling between thiophenol and iodobenzene, it was concluded that 13 mL of water per mmol of thiophenol was the optimal amount (Figure 3).

It is generally accepted that »on-water« processes take place on the water surface or at the organic-aqueous interface, and therefore the reaction outcome is unaffected by dilution [4]. We propose that a higher concentration of the thiol (low dilution) in the latter interphase could have a poisoning effect of the copper catalyst, and in greater dilution the relative amount of the water soluble catalyst is not sufficient to effect the reaction in good yield. Therefore, the optimal amount of water is that which minimizes the influence of both the poisoning effect and the excessive dilution of the catalyst.

References

1 (a) K.R. Bley, R.D. Clark, A. Jahangir, (F. Hoffmann – La Roche AG, CH) Patent WO 2005/005394 A2, **2005** (*Chem. Abstr.* **2005**, *142*, 155951); (b) R. Amorati, M.G. Fumo, S. Menichetti, V. Mugnaini, G.F. Pedulli, *J. Org. Chem.* **2006**, *71*, 6325; (c) G. Liu, J.R. Huth, E.T. Olejniczak, R. Mendoza, P. DeVries, S. Leitza, E.B. Reilly, G.F. Okasinski, S.W. Fesik, T.W. von Geldern, *J. Med. Chem.* **2001**, *44*, 1202; (d) G. Liu, J.T. Link, Z. Pei, E.B. Reilly, S. Leitza, B. Nguyen, K.C. Marsh, G.F. Okasinski, T.W. von Geldern, M. Ormes, K. Fowler, M. Gallatin, *J. Med. Chem.* **2000**, *43*, 4025; (e) A. Pinchart, C. Dallaire, M. Gingras, *Tetrahedron Lett.* **1998**, *39*, 543.

2 (a) S.V. Ley, A.W. Thomas, *Angew. Chem.* **2003**, *115*, 5558; *Angew. Chem. Int. Ed.* **2003**, *42*, 5400; (b) I.P. Beletskaya, A.V. Cheprakov, *Coord. Chem. Rev.* **2004**, *248*, 2337.

3 (a) C. Palomo, M. Oiarbide, R. López, E. Gómez-Bengoa, *Tetrahedron Lett.* **2000**, *41*, 1283; (b) C.G. Bates, R.K. Gujadhur, D. Venkataraman, *Org. Lett.* **2002**, *4*, 2803.

4 (a) S. Narayan, J. Muldoon, M.G. Finn,V.V. Fokin, H.C. Kolb, K.B. Sharpless, *Angew. Chem.* **2005**, *117*, 3339; *Angew. Chem. Int. Ed.* **2005**, *44*, 3275; (b) C.-J. Li, *Chem. Rev.* **2005**, *105*, 3095; (c) C.-J. Li, L. Chen, *Chem. Soc. Rev.* **2006**, *35*, 68.

5 M. Carril, R. SanMartin, I. Tellitu, E. Domínguez, *Org. Lett.* **2006**, *8*, 1467.

6 M. Carril, R. SanMartin, E. Domínguez, I. Tellitu, *Chem. Eur. J.* **2007**, *13*, 5100.

7 H. Takeuchi, T. Hiyama, N. Kamai, H. Oya, *J. Chem. Soc., Perkin Trans. 2* **1997**, 2301.

8 H.J. Cristau, B. Chabaud, A. Chene, H. Christol, *Synthesis* **1981**, 892.

9 For a previous related work, see: S.L. Buchwald, A. Klapars, F.Y. Kwong, E. Streiter, J. Zanon, (Massachusetts Institute of Technology, US), Patent WO 2004/013094, **2004** (*Chem. Abstr.* **2004**, *140*, 181205).

10 P. Barbaro, *Chem. Eur. J.* **2006**, *12*, 5666 and references therein.

4
2-(2′-Anilinyl)-4,4-Dimethyl-2-Oxazoline

Robert A. Gossage

2-Oxazolines, or more correctly 4,5-dihydrooxazoles, are a class of heterocyclic compounds first produced synthetically by Rudolf Andreasch (Karl-Franzens-Universität Graz) in 1884 [1]. The system, which was later correctly identified [2] by Sigmund Gabriel (I. Berliner Universitäts-Laboratorium), consists of a five-membered ring containing a single oxygen atom and a sp^2-hybridized nitrogen atom, the latter being connected via a double bond to a carbon atom linking the two heteroatom centers (Scheme 1: **1**). Some fifty years later, the first naturally occurring oxazoline, produced by the western Canadian weed *Conringia orientalis* (hare's ear), was isolated [3] by Canadian researcher Clarence Hopkins (National Research Laboratories, Ottawa). Since that time many natural and synthetic oxazoline compounds have been reported [4] and this class of heterocycles is now used routinely in medicinal chemistry [5], directed *ortho*-metallation (»DoM«) reactions [6], transition metal and main group coordination chemistry [7], stoichiometric [8] and catalytic regio- and/or enantio-selective C-C bond forming processes [9], and is employed as monomer units for the synthesis of a variety of polymeric materials[10]. In 1937, Leffler and Adams (University of Illinois) described the synthesis of the parent of the title material 2-(2′-anilinyl)-2-oxazoline (**2**) and investigated its properties as an anesthetic agent [11]. Their four-step synthesis included the use of corrosive PBr_3, a toxic thionyl chloride-mediated synthesis of 2-nitrobenzoyl chloride and a reduction step. Related materials, such as the focus of this chapter (**3**) [12], can now be recognized as masked anthranilic acid in which the carboxylic acid functionality has been converted into an oxazoline ring. The 4,4-dimethyl-2-oxazoline heterocycle is now well known as a robust protecting group for such carboxylic acids [4, 13]. The single-step synthesis of **3** detailed here uses the very inexpensive reagent isatoic anhydride (**4**) [14] in combination with an amino-alcohol and includes an atom economic zinc halide-catalyzed [15] oxazoline ring formation protocol. In addition, chromatographic isolation of the title material is avoided.

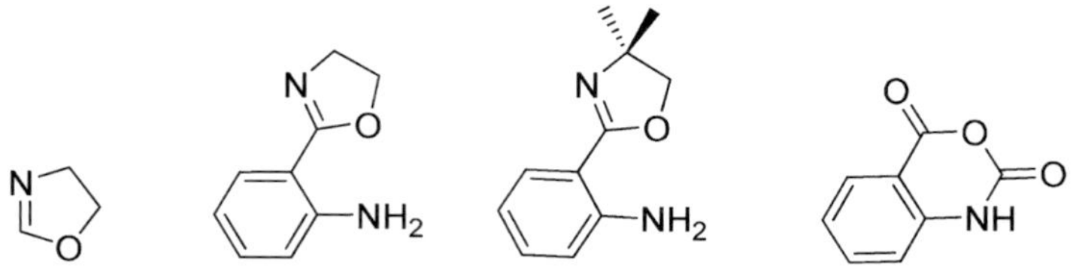

Scheme 1

Preparation of 2-(2′-anilinyl)-4,4-dimethyl-2-oxazoline

Apparatus   A 500 mL three-necked round-bottomed flask with the two side arms equipped with glass stoppers or rubber septa, a magnetic stirrer with heating capability, a suitable reflux condenser topped with a $CaCl_2$ drying tube, a large magnetic stirring bar, a large filter funnel, filter paper (Whatman Grade 202 or equivalent), a 250 mL round-bottomed flask or a 250 mL evaporating dish, one 600 mL beaker and two 500 mL single necked round-bottomed flasks, wooden boiling sticks, a steam bath apparatus, a rotary evaporator, safety glasses, laboratory coat, and protective gloves.

Chemicals   Isatoic anhydride, 2-amino-2-methyl-1-propanol, reagent grade chlorobenzene (predrying or distillation is not required), anhydrous zinc bromide, petroleum ether (90-120 °C b.p. range), diethyl ether.

Attention!   Safety glasses, a laboratory coat and protective gloves must be used at all times.

Caution!   Care should be taken to avoid inhalation of chlorobenzene, petroleum ether, diethyl ether and 2-amino-2-methyl-1-propanol because of their toxicity and volatile nature. These solvents are also flammable. Isatoic anhydride and $ZnBr_2$ are listed as irritants, and hence contact of these materials with the skin should be avoided at all times. All reactions should be carried out in a well-ventilated fume hood.

Procedure   A 500 mL three necked round-bottomed flask is connected to a reflux condenser and the two side arms are closed with glass stoppers. A large magnetic stirring bar is added. The flask is then charged through one of the side arms with 98% isatoic anhydride (10.0 g,

60.1 mmol) and chlorobenzene (~130 mL); vigorous stirring, necessary for optimal yield, is initiated. A 1.41 g quantity of $ZnBr_2$ (6.3 mmol) is added to the mixture which is subsequently warmed to a temperature of about 40 °C. To the warm suspension is added 2-amino-2-methyl-1-propanol (6.0 g, 67 mmol); if necessary any viscous amine can be washed into the reaction mixture with additional chlorobenzene (2 × 10 mL). The resulting combination is then stirred at reflux temperature for a minimum of 12 h. During the initial stages of heating the suspension begins to effervesce; the mixture also darkens slightly in color over the first few hours. Upon conclusion of the period of reflux, the reaction vessel is cooled to room temperature and the mixture filtered (filter paper) to remove any brown colored (insoluble) by-products [16]. Volatile components of this mixture are now removed by either (i) rotary evaporation (following transfer of the organic layer to a 250 mL round-bottomed flask) or (ii) by transferring the solution to a 250 mL evaporating dish and then placing it in a well-ventilated fume hood overnight. The light cream-colored solid obtained in this way (~9 g) is then extracted with 80 mL of diethyl ether and then filtered into a 600 mL beaker containing 300 mL of petroleum ether (90–120 °C b.p. range) and a wooden boiling stick. The milky suspension is heated on a steam bath to a temperature of about 60 °C for a few minutes, followed by hot filtration (filter paper) of the extract into a 500 mL round-bottomed flask. This is set aside to cool to room temperature. The remaining residue from the above is further extracted a second time with a 200 mL portion of hot (80 °C) petroleum ether and filtered as above. The organic extracts yield a white powder on rotary evaporation (alternatively, the extracts can be allowed to slowly evaporate in a large evaporation dish which is placed in a fume hood). The combined yield of the product, 2-(2′-anilinyl)-4,4-dimethyl-2-oxazoline, is 7.10 g (62%). This powder is of sufficient purity for further use; however, the compound can be further recrystallized from hot 80% aq. EtOH (~1 g per 10 mL of solvent mixture) followed by cooling to −10 °C and isolation of the solid by filtration. This process leads to the formation of needle-like colorless plates (recovered: 69%).

Characterization data: m.p. 103–105 °C (uncorrected). ^{1}H NMR (CDCl$_3$): δ 1.38 (s, 6H, CH_3), 4.00 (s, 2H, CH_2), 6.08 (s, br, 2H, NH_2), 6.67 (m, 2H, ArH), 7.21 (m, 1H, ArH), 7.71 (m, 1H, ArH) ppm. ^{13}C{^{1}H} NMR (CDCl$_3$): δ 28.8, 67.8, 77.4, 109.3, 115.6, 116.0, 129.5,

131.9, 148.5, 162.1 ppm. IR (KBr, cm^{-1}): $\bar{v}$ 3428 (m), 3270 (w), 1625 (vs), 1480 (s), 1304 (s), 1045 (vs), 751 (m).

Waste Disposal   Any solid materials filtered off at the conclusion of the reflux period (and subsequent cooling) should be treated as »Halogenated Solid Chemical Waste« due to the possible retention of traces of reaction solvent and ZnBr$_2$ that may be present. These solids should be discarded or destroyed as required by your local environmental regulations. The chlorobenzene (reaction solvent) can be reused and redistilled for future use by recovering this solvent from the rotary evaporator collection flask. Alternatively, this material should be collected and destroyed or disposed of properly and is classified as »Halogenated Solvent Waste«. Residual solvents used in the recrystallization steps should be collected and disposed of as »Nonhalogenated Solvent Waste«.

Explanation

The title compound (alternatively referred to under the names 2-(2'-anilinyl)-4,4-dimethyl-4,5-dihydrooxazole or 2-(4,4-dimethyl-4,5-dihydro-oxazol-2-yl)-aniline) and its congeners have been employed for the attachment of the oxazoline fragment to solid inorganic supports [17], Au nanoparticles [18], as a monomer in polymer synthesis [19], in coordination chemistry [20] and for the design of larger ligand frameworks [21]. This method provides the title material using inexpensive reagents and a readily available Lewis acid catalyst [22]. In addition, chromatographic separation is avoided while material of high purity is still provided.

Acknowledgments

The author is indebted to the support of Research Corporation and NSERC (Canada); Prof. N. D. Jones, (the late) C. R. Eisnor, A. A. Deshpande, and A. R. Boyd are thanked for their assistance with this manuscript.

References

1 R. Andreasch, *Monatsh. Chem.* **1884**, *5*, 33.
2 S. Gabriel, *Chem. Br.* **1889**, *22*, 1139.
3 C.Y. Hopkins, *Can. J. Res. B: Chem. Sci.* **1938**, *16B*, 341.
4 (a) R.H. Wiley, L.L. Bennett, Jr., *Chem. Rev.* **1949**, *44*, 447; (b) J.A. Frump, *Chem. Rev.* **1971**, *71*, 483; (c) B.E. Maryanoff, *Chemistry of Heterocyclic Compounds: Vol. XLV*, **1986**, Ch. 5; (d) T. Eicher, A. Speicher, *The Chemistry of Heterocycles: Structure, Reactions, Syntheses and Applications*; *2nd ed.*, Wiley-VCH, Weinheim, **2003**.
5 For example: (a) H.R. Onishi, B.A. Pelak, L.S. Gerckens, L.L. Silver, F.M. Kahan, M.-H. Chen, A.A. Patchett, S.M. Galloway, S.A. Hyland, M.S. Anderson, C.R.H. Raetz, *Science* **1996**, *274*, 980; (b) T. Kline, N.H. Andersen, E.A. Harwood, J. Bowman, A. Malanda, S. Endsley, A.L. Erwin, M. Doyle, S. Fong, A.L. Harris, B. Mendelsohn, K. Mdluli, C.R.H. Raetz, C.K. Stover, P.R. White, A. Yabannavar, S. Zhu, *J. Med. Chem.* **2002**, *45*, 3112; (c) R.U. Kadam, A. Chavan, N. Roy, *Bioorg. Med. Chem. Lett.* **2007**, *17*, 861; (d) L. Fan, E. Lobkovsky, B. Ganem, *Org. Lett.* **2007**, *9*, 2015; (e) C. Camoutsis, *J. Heterocyclic Chem.* **1996**, *33*, 539.
6 (a) H.W. Gschwend, A. Hamdan, *J. Org. Chem.* **1975**, *40*, 2008; (b) A.I. Meyers, E.D. Mihelich, *J. Org. Chem.* **1975**, *40*, 3159; (c) L. Christiaens, A. Luxen, M. Evers, Ph. Thibaut, M. Mbuyi, A. Welter, *Chem. Scripta* **1984**, *24*, 178; (d) E. Wehman, G. van Koten, J.T.B.H. Jastrzebski, M.A. Rotteveel, C.H. Stam, *Organometallics* **1988**, *7*, 1477; (e) T. Sammakia, H.A. Latham, D.R. Schaad, *J. Org. Chem.* **1995**, *60*, 10; (f) L. Green, B. Chauder, V. Snieckus, *J. Heterocyclic Chem.* **1999**, *36*, 1453; (g) A. Scott, *Aust. J. Chem.* **2003**, *56*, 953.
7 (a) M. Gómez, G. Muller, M. Rocamora, *Coord. Chem. Rev.* **1999**, *193–195*, 769; (b) A. Decken, C.R. Eisnor, R.A. Gossage, S.M. Jackson, *Inorg. Chim. Acta* **2006**, *359*, 1743; (c) C.A. Caputo, F.d.S. Carneiro, M.C. Jennings, N.D. Jones, *Can. J. Chem.* **2007**, *85*, 85; (d) D.J. Berg, C. Zhou, T.M. Barclay, X. Fei, S. Feng, K.A. Ogilvie, R.A. Gossage, B. Twamley, M. Wood, *Can. J. Chem.* **2005**, *83*, 449; (e) R.A. Gossage, H.A. Jenkins, P.N. Yadav, *Tetrahedron Lett.* **2004**, *45*, 7689 and **2005**, *46*,

5243; (f) T.M. Barclay, I. del Río, R.A. Gossage, S.M. Jackson, *Can. J. Chem.* **2003**, *81*, 1482.
8 (a) A.I. Meyers, R.A. Gabel, *J. Org. Chem.* **1977**, *42*, 2653; (b) A.I. Meyers, *Acc. Chem. Res.* **1978**, *11*, 375; (c) A.I. Meyers, H.A. Hanagan, L.M. Trefonas, R.J. Baker, *Tetrahedron* **1983**, *39*, 1991; (d) M. Reuman, A.I. Meyers, *Tetrahedron* **1985**, *41*, 837; (e) P.C. Wong, T.B. Barnes, A.T. Chiu, D.D. Christ, J.V. Duncia, W.F. Herblin, P.B.M.W.M. Timmermans, *Cardiovasc. Drug Rev.* **1991**, *9*, 317; (f) A.I. Meyers, *J. Heterocyclic Chem.* **1998**, *35*, 991; (g) Y. Langlois, *Curr. Org. Chem.* **1998**, *2*, 1; (h) A.I. Meyers, *J. Org. Chem.* **2005**, *70*, 6137.
9 (a) C.A. Caputo, N.D. Jones, *Dalton Trans.* **2007**, *4627*; (b) R. Chincilla, C. Najera, *Chem. Rev.* **2007**, *107*, 874; (c) A. Pfaltz, W.J. Drury III, *Proc. Nat. Acad. Sci. U.S.A.* **2004**, *101*, 5723; (d) H.A. McManus, P.J. Guiry, *Chem. Rev.*, **2004**, *104*, 4151; (e) G. Desimoni, G. Faita, P. Quadrelli, *Chem. Rev.* **2003**, *103*, 3119; (f) F. Fache, E. Schulz, M.L. Tommasino, M. Lemaire, *Chem. Rev.* **2000**, *100*, 2159; (g) C.R. Eisnor, R.A. Gossage, P.N. Yadav, *Tetrahedron* **2006**, *62*, 3395; (h) G.G. Cross, C.R. Eisnor, R.A. Gossage, H.A. Jenkins, *Tetrahedron Lett.* **2006**, *47*, 2245.
10 (a) B.M. Culbertson, *Prog. Polym. Sci.* **2002**, *27*, 579; (b) K. Aoi, M. Okada, *Prog. Polym. Sci.* **1996**, *21*, 151; (c) S. Kobayashi, *Prog. Polym. Sci.* **1990**, *15*, 751.
11 M.T. Leffler, R. Adams, *J. Am. Chem. Soc.* **1937**, *59*, 2252.
12 (a) G.M. Coppola, G.E. Hardtmann, *Synthesis* **1980**, 63; (b) J.N. Reed, V. Snieckus, *Tetrahedron Lett.* **1983**, *24*, 3795; (c) K.M. Button, R.A. Gossage, *J. Heterocyclic Chem.* **2003**, *40*, 513; (d) D.A. Hunt, *Org. Prep. Proceed. Int.* **2007**, *39*, 93; (e) T. Ichiyanagi, M. Shimizu, T. Fujisawa, *J. Org. Chem.* **1997**, *62*, 7937, *Tetrahedron Lett.* **1995**, *36*, 5031.
13 G. Dobson, W.W. Christie, *Trends Anal. Chem.* **1996**, *15*, 130.
14 M.-G.A. Shvekhgeimer, *Khim. Geterosiklicheskikh Soed.* **2001**, *37*, 435, *Chem. Heterocyclic Compds.* **2001**, *37*, 385.
15 (a) H. Witte, W. Seeliger, *Liebigs Ann. Chem.* **1974**, 996; (b) K.M. Button, R.A. Gossage,

R.K.R. Phillips, *Synth. Commun.* **2002**, *32*, 363.

16 The fate of isatoic anhydride upon heating has been investigated; see: (a) F.M. Menger, H.B. Kaiserman, *J. Org. Chem.* **1987**, *52*, 315; (b) S.-J. Chiu, C.-H. Chou, *Tetrahedron Lett.* **1999**, *40*, 9271.

17 A.S. Gajare, N.S. Shaikh, G.K. Jnaneshwara, V.H. Deshpande, *J. Chem. Soc. Perkin Trans. 1* **2000**, 999.

18 J.T. Banks, K.M. Button, R.A. Gossage, T.D. Hamilton, K.E. Kershaw, *Heterocycles* **2001**, *55*, 2251.

19 L. Jiang, H. Ni, *Polym. Bull.* **2004**, *52*, 1.

20 For example: (a) M. Gómez, S. Jansat, G. Muller, G. Aullon, M.A. Maestro, *Eur. J. Inorg. Chem.* **2005**, 4341; (b) J.A. Cabeza, I. da Silva, I. del Río, R.A. Gossage, L. Martínez-Méndez, D. Miguel, *J. Organomet. Chem.* **2007**, *692*, 4346; (c) J.A. Cabeza, I. da Silva, I. del Río, R.A. Gossage, D. Miguel, M. Suárez, *Dalton Trans.* **2006**, 2450; (d) J. Castro, S. Cabaleiro, P. Perez-Lourido, J. Romero, J.A. García-Vázquez, A. Sousa, *Polyhedron* **2001**, *20*, 2329.

21 (a) G.C. Hargaden, H. Müller-Bunz, P.J. Guiry, *Eur. J. Org. Chem.* **2007**, 4235; (b) F.T. Luo, V.K. Ravi, C.H. Xue, *Tetrahedron* **2006**, *62*, 9365; (c) A. Decken, R.A. Gossage, P.N. Yadav, *Can. J. Chem.* **2005**, *83*, 1185; (d) K.-M. Wu, C.-A. Huang, K.-F. Peng, C.-T. Chen, *Tetrahedron* **2005**, *61*, 9679; (e) A.Yu. Rulev, L.I. Larina, Y.A. Chuvashev, V.V. Novokshonov, *Mendeleev Commun.* **2005**, 128; (f) N. End, L. Macko, M. Zehnder, A. Pfaltz, *Chem. Eur. J.* **1998**, *4*, 818; (g) N. End, A. Pfaltz, *Chem. Commun.* **1998**, 589; (h) A. Pfaltz, *J. Heterocyclic Chem.* **1999**, *36*, 1437; (i) M. Gómez, G. Muller, D. Panyella, M. Rocamora, E. Duñach, S. Olivero, J.-C. Clinet, *Organometallics* **1997**, *16*, 5900; (j) J. Sedelmeier, T. Hammerer, C. Bolm, *Org. Lett.* **2008**, *10*, 917; (k) S. Doherty, J.G. Knight, A. McRae, R.W. Harrington, W. Clegg, *Eur. J. Org. Chem.* **2008**, 1759.

22 R.A. Gossage, *Curr. Org. Chem.* **2006**, *10*, 923.

5

A Carbene Transfer Agent

John P. Canal, Taramatee Ramnial and Jason A. C. Clyburne

In 2005, the Nobel Prize in Chemistry was awarded to Chauvin, Grubbs and Schrock for the development of the olefin-metathesis reaction in organic chemistry [1–3]. Olefin-metathesis, which is a reaction that allows for the exchange of olefin groups, is a powerful tool in organic chemistry and can produce a multitude of compounds [4]. As noted in the Nobel Prize citation, olefin-metathesis leads to »Greener« synthesis through the reduction of hazardous byproducts of the chemical reaction. Advances in olefin-metathesis paralleled the discovery of new catalysts by both Grubbs and Schrock [5]. High catalytic activity was achieved with Grubbs« »Second Generation« catalyst (1) (Scheme 1), a ruthenium catalyst which employs a N-heterocyclic carbene (NHC) ligand [3]. Carbenes, which have been the subject of many review articles [6–11], have a divalent carbon center and are often key ligands in a wide variety of catalysts 2-5 (Scheme 1) [10, 11]. Examples of important catalysts include those for reactions such as aryl aminations [7, 12, 13], hydrosilylations [14], and Suzuki coupling [7, 12, 13].

Almost 40 years ago, Öfele and Wanzlick independently published the first preparation and structure of a metal complex containing a NHC ligand [15–17]. In the years that followed, more examples of metal-stabilized carbenes, such as the classic Schrock and Fischer type carbenes, were isolated [18]. The first stable carbene was reported in 1988 by Bertrand, when the singlet carbene, *bis*(diisopropylamino)-phosphino[trimethylsilylcarbene] (4) was isolated [19]. In 1991, the pioneering work of Wanzlick and Öfele was revisited by Arduengo, who isolated the first stable NHC, 1,3-diadamantylimidazol-2-ylidene (3) [20]. Finally, the first persistent triplet state carbene (5) was reported by Tomioka in 1995 [21].

The preparation of stable carbenes often requires tedious and technically demanding procedures because of their high reactivity and sensitivity towards both oxygen and moisture. This sensitivity has inspired research into the area of

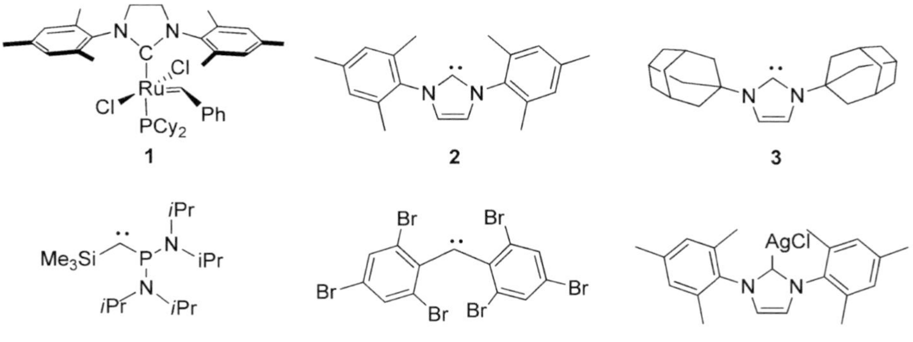

Scheme 1 Grubbs Second Generation catalyst (Ph = phenyl, PCy$_3$ = tricyclo-hexylphosphine) (**1**), NHCs (**2**, **3**), examples of carbenes (*i*Pr = isopropyl) (**4**, **5**) and a carbene transfer agent (**6**).

air-stable molecules that behave as carbene transfer agents [10, 22]. One such compound is the silver(I) imidazolium chloride adduct (**6**) [23, 24]. Despite the central role of NHCs in modern synthetic chemistry, the special requirements needed to work with free carbenes have limited their use in the undergraduate laboratory curriculum. Presented here is a synthetic route to a carbene transfer agent, imidazol-2-ylidene-silver(I) chloride (**6**); an air- and moisture-stable carbene equivalent [23, 24].

Preparation of [(C$_6$Me$_3$H$_2$)$_2$N$_2$(CH)$_2$C]AgCl

Apparatus   10 and 50 mL graduated cylinders, 50 mL and 100 mL round-bottomed flasks, 1 mL syringe, glass stoppers, stir bar, Büchner funnel and filter paper, spatulas, pipettes, analytical balance, weigh paper, sand bath (or metal heating mantle), thermometer, heating and stirring plate, water-cooled condenser, rotary evaporator, NMR tubes and caps, melting point apparatus, safety glasses, laboratory coat, protective gloves.

Chemicals   Deuterated trichloromethane, dichloromethane, diethyl ether, ethanol, 88% formic acid, glyoxal 40 wt% solution in water, 4 mol L^{-1}

hydrochloric acid in 1,4-dioxane, paraformaldehyde, silver(I) oxide, tetrahydrofuran, toluene, 2,4,6-trimethylaniline.

Attention!   Safety glasses and protective gloves must be worn at all times.

Caution!   Some of the chemicals used in this experiment require care in handling, and their use should be restricted to fume hoods: trichloromethane (toxic, irritant), dichloromethane (toxic, irritant), diethyl ether (flammable), ethanol (flammable), formic acid (corrosive), glyoxal (irritant), hydrogen chloride in 1,4-dioxane (flammable, toxic), paraformaldehyde (corrosive), silver(I) oxide (respiratory irritant), tetrahydrofuran (flammable), toluene (flammable, toxic), 2,4,6-trimethylaniline (toxic).

Procedure   The experimental procedure involves three components (Scheme 2). In Step 1, glyoxal-*bis*(2,4,6-trimethylphenyl)imine (**7**) is produced, which is converted into 1,3-*bis*(2,4,6-trimethylphenyl)-imidazolium chloride (**8**) in Step 2. This is the main reactant in Step 3, which ulti-

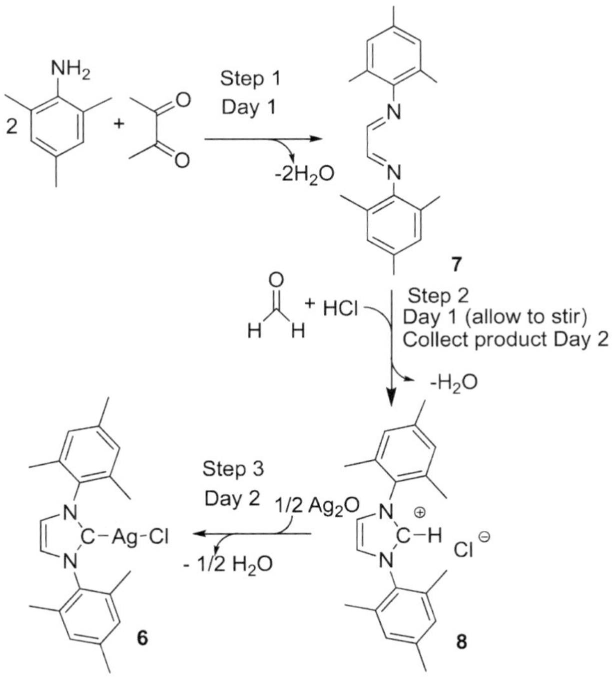

Scheme 2 Full synthetic sequence.

mately yields the carbene transfer agent, imidazol-2-ylidene-silver(I) chloride (**6**).

Step 1: Preparation of glyoxal-bis(2,4,6-trimethylphenyl)imine (7). A 100 mL round-bottomed flask is charged with 10 mL of 2,4,6-trimethyl-aniline (10 mL), 40% glyoxal in water (4 mL) and ethanol (45 mL). A few drops of 88% formic acid are added to initiate precipitation, which is complete after 1 h. The yellow crystals of **7** are collected by vacuum filtration and washed first with cold ethanol (2 × 7 mL), then cold di-ethyl ether (7 mL). Yield 6.76 g (65%).

Characterization data for 7: m.p. 157–160 °C. ^{1}H NMR (400 MHz, CDCl$_3$): δ 2.16 (s, 12 H, *ortho*-CH$_3$), 2.29 (s, 6 H, *para*-CH$_3$), 6.91 (s, 4 H, *meta*-H), 8.10 (s, 2 H, CH) ppm.

Step 2: Preparation of 1,3-bis(2,4,6-trimethylphenyl)-imidazolium chloride (8). A 50 mL round-bottomed flask is charged with glyoxal-bis(2,4,6-trimethylphenyl)imine (1.50 g, 5.13 mmol), paraformalde-hyde (0.15 g, 5.00 mmol) and toluene (30 mL) and capped with a wa-ter-cooled condenser. The stirred solution is heated to 100 °C for 15 min and then allowed to cool to 40 °C. Once cooled, 4 mol L^{-1} HCl in dioxane (1.0 mL) is added drop by drop and the solution is left to stir for one week. (A minimum of three days is required. The stir-ring of the solution is essential to obtain a product, but short inter-ruptions (1–2 h), which allow the stir plates to be used by other stu-dents, do not appear to have a negative effect.) The precipitate is col-lected by vacuum filtration and washed with cold tetrahydrofuran (2 × 5 mL). Yield 1.11 g (65%).

Characterization data for 8: m.p. 349–352 °C. ^{1}H NMR (400 MHz, CDCl$_3$): δ 2.16 (s, 12 H, *ortho*-CH$_3$), 2.32 (2, 6 H, *para*-CH$_3$), 7.01 (s, 4H, *meta*-H), 7.62 (d, 2 H, im-H), 9.72 (s, 1 H, im-H) ppm.

Step 3: Preparation of imidazol-2-ylidene-silver(I) chloride (6). A 100 mL round-bottomed flask is charged with 1,3-*bis*(2,4,6-tri-methylphenyl)-imidazolium chloride (0.50 g, 1.46 mmol) and sil-ver(I) oxide (0.20 g, 8.63 mmol). After adding dichloromethane (30 mL) the solution is stirred and heated to reflux for 2 h. The ex-cess silver(I) oxide is separated by gravity filtration and the solution is concentrated on a rotary evaporator, then cooled to yield a crop of clear crystals of the imidazol-2-ylidene-silver(I) chloride. Yield 0.32 g (50%).

Characterization data for 6: m.p. 280–282 °C. ^{1}H NMR (500 MHz, CDCl$_3$): δ 2.07 (s, 12 H, *ortho*-CH$_3$), 2.35 (s, 6 H, *para*-CH$_3$), 6.99 (s, 4H, *meta*-H), 7.13 (d, 2H, im-H) ppm. ^{13}C{^{1}H} NMR (100.61 MHz,

CDCl$_3$): δ 17.8 (s, o-CH$_3$), 21.3 (s, p-CH$_3$), 122.9 (s, NCC), 129.7 (s, Ar-C-3,5), 134.7 (s, Ar-C-2,6), 135.4 (s, Ar-C-1), 139.8 (s, Ar-C-4), 183.6* (d, ^{107}Ag-**C**), 186.1* (d, ^{109}Ag-**C**) ppm (* only observed in ^{13}C enriched sample).

Waste Disposal The excess/unreacted Ag$_2$O isolated by gravity filtration in Step 3 is washed with dichloromethane, oven dried, collected and used in subsequent reactions. The collected dichloromethane is purified by distillation and can be reused. All other organic solvents are collected and disposed of appropriately.

Explanation

Research into the use of air-stable molecules that behave as carbene transfer agents is driven by the high reactivity and sensitivity of NHCs towards oxygen and moisture [25–28]. Carbene transfer agents, such as the silver(I) imidizolium chloride adduct (**6**) [23, 24], avoid the sometimes tedious and technically demanding preparation of free carbenes. The lability of the C–Ag single bond in this molecule, in concert with the high lattice energy released upon formation of AgCl, are the driving forces for the transfer of the carbene to a range of late transition metals (namely: rhodium, iridium, palladium, and gold, Scheme 3) [22].

The lability of the C-Ag bond has also been utilized in the catalytic role of NHC-AgCl complexes. Through its thermal decomposition, active carbene catalysts can be generated. These organocatalysts can be used for reactions such as the transesterification of small molecules [11, 29]. The single bond between the

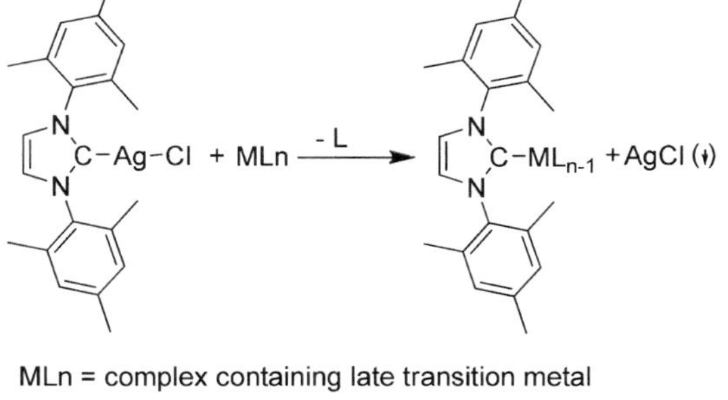

MLn = complex containing late transition metal

Scheme 3 Transfer of a carbene from silver to another late transition metal.

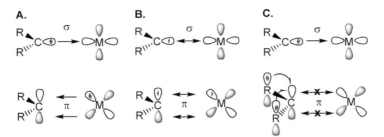

Scheme 4 Representation of the σ and π contribution to the metal-carbon bond of (**A**) Fischer carbene complexes, (**B**) Schrock carbene complexes and (**C**) NHC-metal complexes

metal and the carbenic carbon in NHC-metal complexes differs from the double bond found in Schrock type carbenes. As shown in Schme 4, Schrock type carbenes have a metal-carbon double bond comprising both σ and π components. In NHC-metal complexes, the p orbital on the carbeneic carbon, previously used to π bond to the metal in Fischer and Schrock carbenes, is involved in a back-bonding arrangement with the neighboring N atoms. Therefore, there is primarily a σ component of the metal-carbon bond in NHC-metal complexes, and the average N-C bond lengths are between a double and a single bond [7, 22]. The stability of HNCs results from the delocalization of a lone pair of electrons from N to the carbenic carbon [10].

References

1 Y. Chauvin, *Angew. Chem.* **2006**, *118*, 3824; *Angew. Chem. Int. Ed.* **2006**, *45*, 3741.
2 R.R. Schrock, *Angew. Chem.* **2006**, *118*, 3832; *Angew. Chem. Int. Ed.* **2006**, *45*, 3748.
3 R.H. Grubbs, *Angew. Chem.* **2006**, *118*, 3845; *Angew. Chem. Int. Ed.* **2006**, *45*, 3760.
4 C.E. Housecroft, A.G. Sharpe, *Inorganic Chemistry* 3rd ed.; Pearson Education: New York, **2008**.
5 A.M. Rouhi, *Chemical and Engineering News* **2002**, *80(51)*, 29.
6 J.P. Canal, T. Ramnial, D.A. Dickie, J.A.C. Clyburne, *Chem. Commun.* **2006**, 1809.
7 W.A. Herrmann, *Angew. Chem.* **2002**, *114*, 1342; *Angew. Chem. Int. Ed.* **2002**, *41*, 1290.
8 S. Bindu, V. Nair, V. Sreekumar, *Angew. Chem.* **2004**, *116*, 5240; *Angew. Chem. Int. Ed.* **2004**, *43*, 5130.
9 A.H. Cowley, *J. Organomet. Chem,* **2001**, *617*, 105.
10 D. Bourissou, O. Guerret, F.P. Gabbaï, G. Bertrand, *Chem. Rev.* **2000**, *100*, 39.
11 N. Marion, S. Díez-Gonzáles, S.P. Nolan, *Angew. Chem.* **2007**, *119*, 3046; *Angew. Chem. Int. Ed.* **2007**, *46*, 2988.
12 H. Clavier, S.P. Nolan, *Annu. Rep. Prog. Chem. Sect. B.* **2007**, *103*, 193.
13 A.C. Hillier, G.A. Grasa, M.S. Viciu, H.M. Lee, C. Yang, S.P. Nolan, *J. Organomet. Chem.* **2002**, *653*, 69.
14 V. Dragutan, I. Dragutan, L. Delaude, A. Demonceau, *Coord. Chem. Rev.* **2007**, *251*, 765.
15 K. Öfele, *J. Organomet. Chem.* **1968**, *12*, P42.
16 H.W. Wanzlick, H.J. Schönherr, *Angew. Chem.* **1968**, *80*, 154; *Angew. Chem. Int. Ed. Engl.* **1968**, *7*, 141.

17 P. Luger, G. Ruban, *Acta Crytallogr. Sect. B,* **1971**, *27*, 2276.

18 C. Elsenbroich, A. Salzer, *Organometallics: A Concise Introduction* 2nd ed.; VCH: New York, 1992.

19 A. Igau, H. Grützmacher, A. Baceiredo, G. Bertrand, *J. Am. Chem. Soc.* **1988**, *110*, 6463.

20 A.J. Arduengo III, R.L. Harlow, M. Kline, *J. Am. Chem. Soc.* **1991**, *113*, 361.

21 H. Tomioka, T. Watanabe, K. Hirai, K. Furukawa, T. Takui, K. Itoh, *J. Am. Chem. Soc.* **1995**, *117*, 6376.

22 J.C. Garrison, W.J. Youngs, *Chem. Rev.* **2005**, 105, 3978.

23 J.P. Canal, T. Ramnial, L.D. Langlois, C.D. Abernethy, J.A.C. Clyburne, *J. Chem. Educ.* **2008**, *85*, 416.

24 T. Ramnial, C.D. Abernethy, M.D. Spicer, I.D. Mackenzie, I.D. Gay, J.A.C. Clyburne, *Inorg. Chem.* **2003**, *42*, 1391.

25 H.M.J. Wang, I.J.B. Lin, *Organometallics* **1998**, *17*, 972.

26 A.R. Chianese, X. Li, M. C. Janzen, J.W. Faller, R.H. Crabtree, *Organometallics* **2003**, *22*, 1663.

27 M.D. Sanderson, J.W. Kamplain, C.W. Bielawski, *J. Am. Chem. Soc.* **2006**, *128*, 16514.

28 A.A.D. Tulloch, A.A. Danopoulos, G.J. Tizzard, S.J. Coles, M.B. Hursthouse, R.S. Hay-Motherwell, W.B. Motherwell, *Chem. Commun.* **2001**, 1270.

29 A.C. Sentman, S. Csihony, R. M. Waymouth, J. L. Hedrick, *J. Org. Chem.* **2005**, *70*, 2391.

6

Enantioselective Organocatalytic Synthesis of (R)- and (S)-3,3,3-Trifluoro-2-(Indol-3-yl)-2-Hydroxyl-Propionic Acid Ethyl Esters

Mohammed Abid and Béla Török

The emergence of chiral drugs highlights the importance of asymmetric synthesis [1]. Chiral catalysis is one of the most attractive methods of asymmetric synthesis. It is more likely to satisfy the recent trends of tightening environmental and safety restrictions than any other method and is in compliance with the principles of green chemistry. Organocatalysis [2] is one of the most attractive ways of carrying out asymmetric reactions. Asymmetric catalysis had long been dominated by chiral transition metal complexes and enzymes. In fact, the first practical enantioselective catalytic reactions were developed in the 1960s, and since then these catalysts have ruled the development of new asymmetric reactions. These efforts were honored by the Nobel Prize in chemistry (2001) [3]. In contrast, it has long been known that simple organic molecules were also able to catalyze reactions, as the first chiral catalysts were such compounds. In 1912 Breding described an organocatalytic cyanhydrin synthesis with moderate enantioselectivity. In the 1960s Pracejus was the first to obtain significant enantioselectivity with organic catalysts. A decade later chemists at Roche and Schering described the first proline-catalyzed enantioselective aldol reaction. Natural compounds such as alkaloids and aminoacids dominated this field, that was attracting little interest. About 10 years ago, an explosion-like development began in the field of asymmetric organocatalysis, a term coined by MacMillan in 2000 [4]. It enjoys very significant attention due, in part, to a pressing need to carry out chiral reactions in an environmentally benign way. The high stability, easy manageability and metal-free nature of these catalysts provide them major advantages in the development of new, *green* technologies.

The particular organocatalysts that we apply in the current synthesis belong to the group of the well-known cinchona alkaloids, which are natural compounds [5]. They are found in the bark of the *cinchona* tree, that is indigenous at higher elevations in the tropical Andes. The name itself was given by Linnaeus in 1742 and originated from the (misspelt) name of the Count of Chinchón, who was the Spanish Viceroy in Peru, from 1628 to 1639. According to the legend, his wife became gravely ill with malaria, and after European medicines failed, her physician turned to the traditional medicine of the natives, which miraculously cured the Countess. In fact, the name of the compound responsible for the miracle, *quinine*, originated from the native Peruvian name of the cinchona tree (kina tree, or in Spanish – quina) [5]. Later, Jesuits brought the powder of the back to Europe, where malaria was widespread in those times. It has also performed miracles in Europe: the Papal Conclave in 1655 was the first one without a death from malaria among the cardinals. However, the *Jesuits powder* as it was called was not for everyone. Oliver Cromwell, a devout protestant and the leader of the English Revolution refused to take it when became ill with malaria, and finally died of it in 1658 [6].

Quinine (Figure 1) was first isolated by Pelletier and Caventou in 1820 [7]. Many attempts were made to accomplish its synthesis. The most notable failure is probably Perkin's synthesis. Due to the very limited knowledge of the times, Perkin tried to combine 2 allyltoluidine molecules with 3 oxygen atoms to get quinine (a simple mathematical addition of the C, N, O, and H atoms). He did not obtain quinine, but the compound he isolated, mauveine, was also remarkable. With its beautiful, intense purple color, it became the first synthetic dye and the foundation of the vast synthetic dye industry. An important further attempt at the synthesis of quinine was by Woodward and Doering in 1944. Although, their title suggested the total synthesis of quinine, they only synthesized a precursor that was claimed earlier to be the precursor to quinine. Finally, the synthesis was completed in 2001 by Stork [8], and little later via a different approach by Jacobsen [9]. Figure 1 shows two pairs of the most common

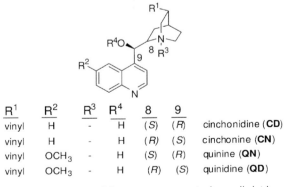

R^1	R^2	R^3	R^4	8	9	
vinyl	H	-	H	(*S*)	(*R*)	cinchonidine (**CD**)
vinyl	H	-	H	(*R*)	(*S*)	cinchonine (**CN**)
vinyl	OCH$_3$	-	H	(*S*)	(*R*)	quinine (**QN**)
vinyl	OCH$_3$	-	H	(*R*)	(*S*)	quinidine (**QD**)

Figure 1 Structures of the most common cinchona alkaloids.

cinchona alkaloids (**CD/QN** and **CN/QD**) that are usually called pseudoenantiomers, and in most reactions provide the opposite enantiomers of the product [7]. Cinchona catalysts have the ultimate advantage over metal complex-based catalysts; they are stable and easy to work with and remove from the product, and most show little toxicity.

While fluorine is the most abundant halogen in rocks of the Earth, organofluorine compounds are extremely rare in nature [10]. Although biologically active organofluorine compounds have been known since the 1940s, the lack of convenient and effective fluorinating agents strongly inhibited the development of this field. The real breakthrough occurred when Fried and Sabo first reported that 9-fluorohydrocortisone acetate possessed about 10 times higher biological activity than its nonfluorinated counterpart [11]. Since then, a steady stream of new, novel fluorinated drugs has been produced, bringing significant advances into drug development, including steroidal and nonsteroidal anti-inflammatory drugs, anticancer agents, antiviral agents and central nervous system drugs (e.g. Prozac (antidepressant), Desflurane (inhalation anesthetic), Casodex (anticancer agent), Befloxatone (antidepressant), Diflucan (antifungal agent) and Efavirenz (anti HIV agent)) [12]. The fact that about 20% of the recently registered drugs and more than 40% of the agrochemicals contain fluorine makes fluorine the most popular heteroatom after nitrogen in life sciences.

The understanding of fundamental biochemical mechanisms coupled with the knowledge of physicochemical properties of fluorine and organofluorine compounds provide sufficient tools for rational drug design [10]. There are several features that make fluorinated organic compounds unique, such as the small van der Waals radius (1.35 Å), the stability of the C-F bond and its electronic effect, the lipophilicity of organofluorine compounds, and fluorine's exceptional hydrogen bond-forming ability. Based on these features, the design and synthesis of fluoroorganic compounds have become an important focus of drug development. The large number of drugs and agrochemicals available on the market clearly illustrates this trend.

We have recently synthesized the target compound in both racemic and chiral forms [13]. This compound and its related derivatives are currently considered as potential fibrillogenesis inhibitors against Alzheimer's disease [14]. A simple Friedel–Crafts hydroxyalkylation will be applied to synthesize this compound using cinchonidine (**CD**) and cinchonine (**CN**) as catalysts.

Synthesis of ethyl (S)-3,3,3-trifluoro-2-(indol-3-yl)-2-hydroxyl-propionate

Apparatus   5 mL glass vial with a screw cap, magnetic stirrer, syringe (500 µL), salt-ice cooling bath, safety glasses, laboratory coat, protective gloves.

Chemicals   Indole, ethyl 3,3,3-trifluoropyruvate, cinchonine (**CN**), cinchonidine (**CD**), quinidine (**QD**), diethyl ether, K-10 montmorillonite (solid acid).

Attention!   Safety glasses and protective gloves must be worn at all times.

Caution!   Because of their toxicity and volatility, care should be taken to avoid inhalation of diethyl ether and ethyl 3,3,3-trifluoropyruvate or contact of their solution with the skin. All reactions should be carried out in a well-ventilated hood.

Procedure   Indole (88 mg, 0.75 mmol) and cinchonidine (**CD**, 11 mg, 0.0375 mmol) are placed in a glass reaction vial and mixed with 3 mL Et_2O. The mixture is stirred at room temperature for 30 min to give a turbid solution. Ethyl 3,3,3-trifluoropyruvate (150 µL, 1.125 mmol) is then added to the above mixture to obtain a clear liquid. The resulting mixture is further stirred at room temperature for 35 min and the progress of the reaction is monitored by thin-layer chromatography (CH_2Cl_2 as eluent). After the reaction is complete, the solvent and excess ethyl trifluoropyruvate are removed by evaporation. The mixture is then dissolved in ether and the cinchonidine is removed by a treatment with 500 mg of K-10 (a solid acid). After the treatment, cinchonidine-K-10 complex is removed by filtration and the solvent is evaporated. A colorless solid is obtained in 98% yield.

Note   Following the above methodology, but using cinchonine (**CN**) as catalyst (the molecular mass is identical with that of **CD**, therefore the same amount, 11 mg, 0.0375 mmol, should be used) the (R)-enantiomer of the product will be obtained in 94% yield.

   Characterization data: (Note that all physical characteristics such as melting point or spectral data of chiral compounds are identical except for optical rotation) m.p. 68.2–70.3 °C. ^1H NMR: (300.12 MHz,

CDCl$_3$), δ 8.23 (bs, 1H, NH), 7.86 (d, J = 7.50 Hz, 1H, Ar), 7.29 (d, J = 2.70 Hz, 1H, Ar), 7.29–7.10 (m, 3H, Ar), 4.47–4.17 (m, 2H, CH$_2$), 4.17 (s, 1H, OH), 1.30 (t, J = 7.20 Hz, 3H, CH$_3$) ppm. ^{13}C NMR: (75.47 MHz, CDCl$_3$), δ 169.3, 136.2, 125.3, 124.9, 124.4, 124.4, 122.5, 121.6, 120.8, 120.4, 111.4, 64.3, 13.8 ppm. ^{19}F NMR: (282.40 MHz, CDCl$_3$, CFCl$_3$-Ref), δ –76.75 (s, 3F) ppm. MS: m/z 287 (M$^+$, 33%), 214 (100%). IR (cm^{-1}): 3308, 1732, 1424, 1225, 778.

Enantiomeric excess determination by HPLC

Product is dissolved in isopropyl alcohol and subjected to HPLC separation under the following conditions:

Column:	Chiralcel OJ-H
Eluent:	Mixture of isopropyl alcohol and *n*-hexane (10:90)
Flow rate:	1 mL min^{-1}
Wavelength:	260 nm
Peak positions:	110 min (93%) (*S*), 127 min (7%) (*R*)

Enantiomeric excess determination by ^{19}F NMR [15]

^{19}F NMR is a particularly simple alternative to HPLC (high-pressure liquid chromatography) separations. As the resonance of the CF$_3$ group is determined (one resonance only), the presence of the several protons in the alkaloid structure or the product itself does not complicate the analysis. ^{19}F NMR analysis is carried out on a 400-MHz superconducting Varian-Inova NMR spectrometer operating at 376 MHz in CDCl$_3$ solvent with CFCl$_3$ as an internal standard. Product (8 mg, 0.023 mmol) and quinidine (**QD**, 22 mg, 0.069 mmol) are mixed with CDCl$_3$ in a NMR sample tube which is later subjected to chiral discrimination. The ^{19}F NMR shows two resonances at –76.388 ppm (93%) for the (*S*)-enantiomer and –76.444 ppm (7%) for the minor (*R*) enantiomer.

Waste Disposal Put the recovered organic solvents into a designated solvent recovery container. The recovered dichloromethane and CDCl$_3$ should be collected separately in a recovery container labeled for halogenated solvents. Put K-10 montmorillonite into a designated container.

Explanation

The Friedel–Crafts alkylation of aromatics with activated carbonyl compounds is one of the most important C-C bond forming reactions. Despite its importance, the number of asymmetric Friedel–Crafts reactions is limited and mostly based on the use of chiral metal complex-based Lewis acids. Cinchona alkaloids are very effective catalysts for the target reaction. Using the pseudoenantiomeric pair (**CD** and **CN**) of the alkaloid as catalysts, they can generate both enantiomers of the product, respectively.

Both **CD** and **CN** provided the enantiomeric products in excellent optical purity and yields. Cinchona alkaloids as catalysts significantly increase the reaction rate to give quantitative yields in 35 min. In contrast, the noncatalyzed racemic reaction takes 24 h to provide similar yields. This observation unambiguously indicates that the enantiodifferentiation is a kinetic phenomenon in this reaction; the reaction yielding the chiral product proceeds with significantly higher rate than the racemic reaction.

The procedure can be extended, if desired, to different substituted indoles and pyrroles. Various indole derivatives have been found to show similar behavior in the above reaction (Scheme 1). However, as a limitation of this approach, 1-substituted indoles always gave good yields but only *racemic* products. Substituents in the 2 position of indole also play an important role; groups with increasing size resulted in gradual decrease in enantioselectivities.

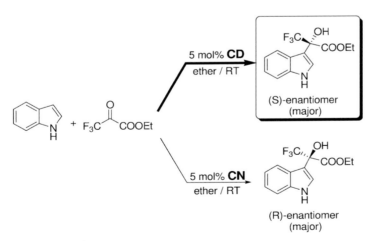

Scheme 1   Hydroxyalkylation of indole with ethyl trifluoropyruvate.

The above organocatalytic approach is a versatile method for the Friedel–Crafts hydroxyalkylation of indoles with ethyl trifluoropyruvate. The major advantages of this process are: clean and fast reactions and the commercially available economic catalysts that can provide both enantiomers of the products.

References

1 (a) R.A. Sheldon, *Chirotechnology: Industrial Synthesis of Optically Active Compounds*, (Marcel Dekker, New York, **1993**); (b) R. Noyori, *Asymmetric Catalysis in Organic Synthesis*, Wiley, New York, **1994**; (c) E.N. Jacobsen, A. Pfaltz, H. Yamamoto (eds.), *Comprehensive Asymmetric Catalysis*, Springer-Verlag, Berlin, **1999**; (d) I. Ojima, *Catalytic Asymmetric Synthesis 2nd ed.*, Wiley-VCH, New York, **2000**; (e) G.-Q. Lin, Y.-M. Li, A.S.-C. Chan, *Principles and Application of Asymmetric Synthesis* Wiley, New York, **2001**.

2 (a) P.I. Dalko, L. Moisan, *Angew. Chem.* **2001**, *113*, 3840; *Angew. Chem. Int. Ed.* **2001**, *40*, 3726; (b) B. List, *Tetrahedron* **2002**, *58*, 5573; (c) K.N. Houk, B. List (eds.) *Acc. Chem. Res.* **2004**, *37*(8th issue) – a special issue on organocatalysis.

3 (a) W.S. Knowles, *Angew. Chem.* **2002**, *114*, 2096; *Angew. Chem. Int. Ed.* **2002**, *41*, 1998; (b) R. Noyori, *Angew. Chem.* **2002**, *114*, 2108; *Angew. Chem. Int. Ed.* **2002**, *41*, 2008; (c) K.B. Sharpless, *Angew. Chem.* **2002**, *114*, 2126; *Angew. Chem. Int. Ed.* **2002**, *41*, 2024; (d) www.nobelprize.org .

4 K.A. Ahrendt, C.J. Borths, D.W.C. MacMillan, *J. Am. Chem. Soc.* **2000**, *122*, 4243.

5 W. Solomon, *The Cinchona Alkaloids*, in *Chemistry of the Alkaloids* (ed. S.W. Pelletier), Van Nostrand Reinhold, New York, Chapter 11. p. 301, **1970**.

6 P. Le Couteur, J. Burreson, *Napoleon's Buttons, 17 Molecules that Changed History*, Tarcher/Penguin, New York, p. 332, **2003**.

7 (a) H. Wynberg, *Top. Stereochem.* **1986**, *16*, 87; K. Kacprzak, J. Gawronski, *Synthesis* **2001**, 961; (b) S.K. Tian, Y. Chen, J. Hang, L. Tang, P. McDaid, L. Deng, *Acc. Chem. Res.* **2004**, *37*, 621.

8 G. Stork, D. Niu, A. Fujimoto, E.R. Koft, J.M. Balkovec, J.R. Tata, G.R. Dake, *J. Am. Chem. Soc.* **2001**, *123*, 3239.

9 I.T. Raheem, S.N. Goodman, E.N. Jacobsen, *J. Am. Chem. Soc.* **2004**, *126*, 706.

10 (a) T. Hiyama, (ed.), *Organofluorine Compounds*, Springer-Verlag, **2000**; (b) P. Kirsch, *Modern Fluoroorganic Chemistry: Synthesis, Reactivity, Applications*, Wiley-VCH, New York, Heidelberg, **2004**.

11 J. Fried, E.F. Sabo, *J. Am. Chem. Soc.* **1954**, *76*, 1455.

12 (a) I. Ojima, J.R. McCarthy, J.T. Welch, (eds.) *Biomedical Frontiers of Fluorine Chemistry*, ACS Symposium Series 639, American Chemical Society, Washington, DC, **1996**; V.A. Soloshonok, (ed.), *Enantiocontrolled Synthesis of Fluoroorganic Compounds: Stereochemical Challenges and Biomedicinal Targets*, Wiley, New York, **1999**; (b) P.V. Ramachandran, (ed.), *Asymmetric Fluoroorganic Chemistry*, ACS Symposium Series 746, American Chemical Society, Washington, DC, **2000**.

13 (a) B. Török, M. Abid, G. London, J. Esquibel, M. Török, S.C. Mhadgut, P. Yan, G.K.S. Prakash, *Angew. Chem.* **2005**, *117*, 3146; *Angew. Chem. Int. Ed.* **2005**, *44*, 3086; (b) M. Abid, B. Török, *Adv. Synth. Catal.* **2005**, *347*, 1797.

14 M. Török, M. Abid, S.C. Mhadgut, B. Török, *Biochemistry* **2006**, *45*, 5377.

15 M. Abid, B. Török, *Tetrahedron: Asymmetry* **2005**, *16*, 1547.

7

Synthesis of a Pyrrole, an Indole, and a Carbazole Derivative by Solid Acid-Catalyzed Electrophilic Annelations

Mohammed Abid and Béla Török

The phenomenon of catalysis is probably almost as old as life on Earth and was in the form of early biocatalysis. Moreover, certain catalytic materials most likely were present at the formation of the early biomolecules. The search for the Philosopher's Stone in medieval times by alchemists can be considered as a quest to find the ultimate catalyst that would turn common materials to gold. Probably Kirchhoff was the first to describe a catalytic reaction in 1812, when he carried out the acid-catalyzed hydrolysis of sugars. A few years later Davy reported the first metal-catalyzed transformation, which he called a *flameless burning*. The word itself and the first definition of catalysis originated from Berzelius in 1836 [1]. Showing its importance, catalysis or its researchers were given continuous recognition over the history of modern science. A long list of Nobel laureates indicates the probably never-ending need for such knowledge (Ostwald – 1909; Sabatier – 1912; Haber – 1918; Bosch – 1931; Langmuir – 1932; Ziegler and Natta – 1963; Knowles, Noyori and Sharpless – 2001, Chauvin, Grubbs and Schrock – 2005, and Ertl – 2007) [2]. Catalysis has also contributed to shaping history. A catalytic reaction (isomerization) in the course of producing and supplying high octane gasoline for the airplanes of the Royal Air Force, probably contributed to the defeat of the Luftwaffe in the aerial battle for Great Britain in the Second World War. With the improved gasoline, the British warplanes just flew faster than the German Messerschmitts (which had the same engine), securing the advantage in confrontations.

In recent years, catalysis has become a core tool for Green Chemistry. Sustainable industrial development is unimaginable without catalytic processes. Catalysis can be practiced in many forms. However, almost undoubtedly, heterogeneous catalysis is the most significant contributor to sustainable developments in the chemical industry. Heterogeneous catalysts can be divided into two major groups: metal- [3] and acid-based[4] catalysts. In the current preparation we will use a solid acid catalyst, K-10 montmorillonite. This material belongs to the group of clay-based materials. Although natural aluminosilicates themselves, clays are significantly different from zeolites, which are probably the most widely used solid acid catalysts. In contrast to the channel and cage structure of zeolites, clays are aluminosilicates with a layered structure similar to that of graphite.

The application of clay catalysts in organic synthesis has a rather extensive history, K-10 and KSF-montmorillonites being the two most commonly applied [4]. Both are synthetic clays produced from natural montmorillonites and are commercially available. K-10 has a substantially higher surface area ($250 \, m^2 \, g^{-1}$) than that of KSF ($10 \, m^2 \, g^{-1}$), and they are comparable in acidity. The intrinsic acidity of natural clays is quite weak. However, when they are treated with mineral acids such as aqueous HCl, dealumination significantly enhances their acidity. The Hammett acidity constant [6] of these materials is $H_0 \approx -8$ [7], which corresponds approximately to the acidity of concentrated nitric acid. In appearance, K-10 is a light brown or yellowish material that is nonhazardous, noncorrosive and easy to handle and work with. Finally, it is also very stable, its shelf life being practically unlimited. Because of its physical nature and polar structure, K-10 absorbs microwave energy and is an excellent catalyst for microwave-assisted organic synthesis (MAOS) [3].

The application of microwave irradiation in organic synthesis was first disclosed in 1986 [8]. Since then the field of microwave-assisted chemical synthesis has gone through an explosive development, nearly 3000 papers having been published in this area. It is used in organic, inorganic, organometallic, polymer and peptide synthesis, and beyond these other applications in biology, geology, and chemical analysis can also be mentioned. Without discussing the physical principles of the microwave effect, we briefly mention that a polar nature of a material is usually advantageous in these reactions. Polar materials (e.g. solvents, reactants or catalysts) are able to absorb microwaves and transform their energy into heat. As this transfer occurs inside the liquid or solid medium (or both), microwave heating is much more effective than conventional convective heating methods. This higher efficiency results in better chemical yields in shorter reaction times. Sometimes it can be dramatic. For example, a condensation reaction of α,α,α-trifluoromethyl ketone with benzaldehyde under conventional conditions (reflux) yields 75% of the product in 168 h (a week!), while using microwave-assisted K-10 catalysis this reaction took place in 30 min and provided a 95% yield [9]. As the reduction of energy consumption is key in designing green processes, microwave synthesis has earned its place among the core technologies of sustainable chemistry.

In the current application we will illustrate the use of both heterogeneous acid catalysis and microwave irradiation in the synthesis of N-containing hetero-cyclic compounds such as pyrrole and its homologous benzenoid derivatives, indole and carbazole [10]. This chemistry has recently been developed in our laboratory and has found many extensions in our synthetic practice [11, 12].

K-10 montmorillonite-catalyzed microwave-assisted synthesis of a pyrrole, an indole, and a carbazole derivative

Apparatus   25 mL round-bottomed flask, magnetic stirrer, vial, microwave reac-tor, 100 and 500 µL syringes, safety glasses, laboratory coat, protec-tive gloves.

Note   The reactions are carried out in a top-loader one-vessel Discover Benchmate (CEM) microwave reactor. The temperature is moni-tored by an IR controller. It is anticipated that the experiments can be carried out in oven-style instruments such as the MARS (CEM) or START (Milestone) systems, which are equipped with fiberoptic temperature controllers.

Chemicals   Aniline, pyrrole, indole, 2,5-hexanedione, dichloromethane, K-10 montmorillonite (solid acid).

Attention!   Safety glasses and protective gloves must be worn at all times.

Caution!   Because of their toxicity and volatility, care should be taken to avoid inhalation of dichloromethane, aniline, pyrrole and indole or skin contact with their solutions. All reactions should be carried out in a well-ventilated hood.

Synthesis of 1-phenyl-2,5-dimethylpyrrole

Procedure   The reaction is carried out in a focused CEM Discover Benchmate microwave reactor, using the open vessel technique. Aniline (91 µL, 1 mmol) and 2,5-hexanedione (176 µL, 1.5 mmol) are placed in a 25 mL round-bottomed flask and dissolved in 10 mL of di-chloromethane; then K-10 montmorillonite (500 mg) is added to the mixture. After 2 min stirring, the solvent is evaporated *in vacuo*. The dry mixture is then transferred into a vial and inserted into the cavity of the microwave reactor. The dry mixture is irradiated at 90 °C for 3 min under atmospheric pressure. The progress of the reaction is monitored by TLC and GCMS. When the reaction is complete the mixture is suspended in CH_2Cl_2. The product is separated from the catalyst by filtration and the solvent is removed *in vacuo*. The crude 1-phenyl-2,5-dimethyl pyrrole can be purified by recrystallization or column chromatography as time allows.

Characterization data: Dark semi solid. Yield 98%. ^1H NMR (399.81 MHz, $CDCl_3$): δ 7.44 (m, 3H, Ar), 7.21 (m, 2H, Ar), 5.91 (s, 2H, Ar), 2.05 (s, 6H, CH_3) ppm. ^{13}C NMR (100.53 MHz, $CDCl_3$): δ 139.2, 129.2, 129.0, 128.4, 127.8, 105.8, 13.2 ppm. MS: *m/z* 171 (M$^+$, 80%), 170 (100%).

Synthesis of 4,6-dimethyl-1H-indole

Procedure

The reaction is carried out in the same system under the same conditions as above. Pyrrole (70 μL, 1 mmol) and 2,5-hexanedione (176 μL, 1.5 mmol) are placed in a 25 mL round-bottomed flask and dissolved in dichloromethane (10 mL); then K-10 montmorillonite (500 mg) is added, and after 2 min stirring the solvent is evaporated. The mixture is then irradiated at 90 °C for 2 min. When the reaction is complete the mixture is treated with CH_2Cl_2. The product is separated from the catalyst by filtration and the solvent is removed *in vacuo*. The crude 4,6-dimethyl-1H-indole can be purified by recrystallization or column chromatography as time allows.

Characterization data: m.p. 48.3–50.1 °C. Yield 98%. ^1H NMR (399.81 MHz, $CDCl_3$): δ 8.05 (bs, 1H, NH), 7.20 (t, J = 2.8 Hz, 1H, Ar), 6.90 (d, J = 7.2 Hz, 1H, Ar), 6.83 (d, J = 7.2 Hz, 1H, Ar), 6.57 (t, J = 2.0 Hz, 1H, Ar), 2.53 (s, 3H, CH_3), 2.46 (s, 3H, CH_3) ppm. ^{13}C NMR (100.53 MHz, $CDCl_3$): δ 135.2, 127.9, 127.4, 123.3, 122.7, 120.1, 117.7, 101.8, 18.7, 16.6 ppm. MS: *m/z* 145 (M$^+$, 100).

Synthesis of 1,4-dimethylcarbazole

Procedure

The reaction is carried out in the same system under same conditions as above. Indole (117 mg, 1 mmol) and 2,5-hexanedione (176 μL, 1.5 mmol) are placed in a 25 mL round-bottomed flask and dissolved in dichloromethane (10 mL); then K-10 montmorillonite (500 mg) is added and after 2 min stirring the solvent is evaporated. The mixture is then irradiated at 90 °C for 12 min. When the reaction is complete the mixture is suspended in CH_2Cl_2. The product is separated from the catalyst by filtration and the solvent is removed *in vacuo*. The crude 1,4-dimethylcarbazole can be purified by recrystallization or column chromatography as desired and as time allows.

Characterization data: m.p. 68.5–70.2 °C. Yield 95%. ^{1}H NMR (399.81 MHz, $CDCl_3$): δ 8.18 (d, $J = 8.0$ Hz, 1H, Ar), 7.97 (bs, 1H, NH), 7.47 (d, $J = 8.0$ Hz, 1H, Ar), 7.41 (dt, $J = 7.2$, 0.8 Hz, 1H, Ar), 7.25 (dt, $J = 7.2$, 0.8 Hz, 1H, Ar), 7.13 (d, $J = 7.2$ Hz, 1H, Ar), 6.94 (d, $J = 7.2$ Hz, 1H, Ar), 2.86 (s, 3H, CH3), 2.53 (s, 3H, CH_3) ppm. ^{13}C NMR (100.53 MHz, $CDCl_3$): δ 139.5, 138.9, 131.0, 126.3, 125.2, 124.7, 122.8, 121.5, 121.1, 119.6, 117.1, 110.6, 20.7, 16.7 ppm. MS: m/z 195 (M$^+$, 100%), 180 (50%), 167 (10%), 115 (5%).

Waste Disposal   Put the recovered dichloromethane in a recovery container labeled for halogenated solvents. Collect the K-10 montmorillonite from the filter and put it into a designated container.

Explanation

All the reactions (Scheme 1) provide practically quantitative yields in very short irradiation times. Synthesis of pyrrole from aniline follows the general Paal-

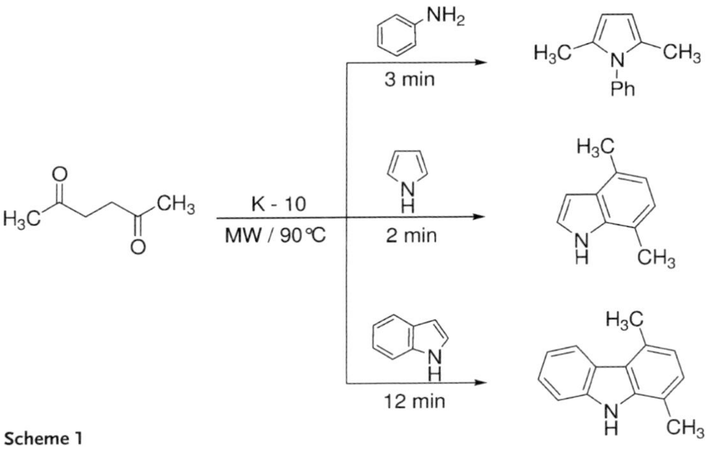

Scheme 1

Knorr-type mechanism [11,12]. The synthesis of indole and carbazole occurs through electrophilic annelation. In a step-wise process, first hydroxyalkylation occurs to form an unstable intermediate. This intermediate immediately undergoes dehydration and loses two water molecules.

The driving force for the easy dehydration step is probably the significant stabilization energy that is provided by the formation of the new aromatic system. The mechanism is briefly illustrated below (Scheme 2).

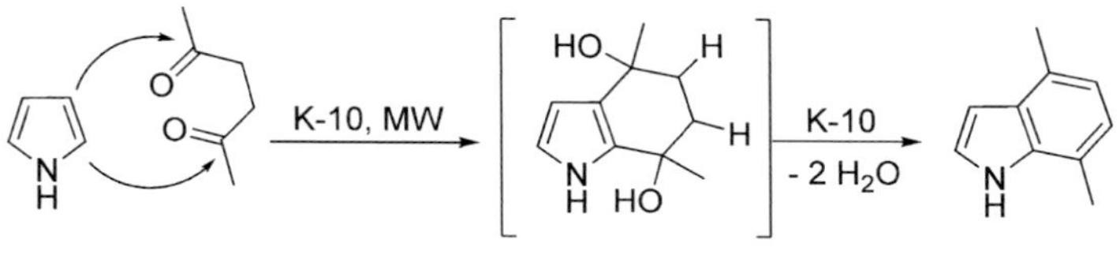

Scheme 2

This method can be extended for the synthesis of substituted pyrroles, indoles and carbazoles. The major advantages of the approach are: high atom economy, solvent-free reaction, solid acid catalysis, very limited energy consumption and waste-free (H_2O is the only byproduct). All these make this process an environmentally benign alternative for the synthesis of these important heterocycles.

References

1 J.K. Smith, History of Catalysis, in *Encyclopedia of Catalysis*, I.T. Horváth (ed.), Wiley, New York, **2003**, Vol. 3. p. 447.

2 www.nobelprize.org ; the official website of the Nobel Foundation.

3 G.V. Smith, F. Notheisz, *Heterogeneous Catalysis in Organic Chemistry*, Academic Press, San Diego, **1999**.

4 B.C. Gates, Catalysis by Solid Acids, in *Encyclopedia of Catalysis*, I.T. Horváth (ed.), Wiley, New York, **2003**, Vol. 1. p. 104.

5 S. Dasgupta, B. Török, Application of Clay Catalysts in Organic Synthesis. A Review. *Org. Prep. Proc. Int.* **2008**, *40*, 1.

6 G.A. Olah, G.K.S. Prakash, J. Sommer, *Superacids*, Wiley, New York, **1985**.

7 M. Balogh, P. Laszlo, *Organic Chemistry Using Clays*, Springer-Verlag, Berlin/Heidelberg, **1993**.

8 A. Loupy, *Microwaves in Organic Synthesis*, Wiley-VCH, **2006**.

9 M. Abid, M. Savolainen, S.M. Landge, J. Hu, G.K.S. Prakash, G.A. Olah, B. Török, *J. Fluorine Chem.* **2007**, *128*, 587.

10 M. Abid, A. Spaeth, B. Török, *Adv. Synth. Catal.* **2006**, *348*, 2191.

11 M. Abid, S. M. Landge, B. Török, *Org. Prep. Proc. Int.* **2006**, *38*, 495.

12 M. Abid, O. De Paolis, B. Török, *Synlett* **2008**, 410.

13 B.K. Banik, S. Samajdar, I. Banik, *J. Org. Chem.* **2004**, *69*, 213.

8

Synthesis of 1,3,5-Triphenylpyrazole by a Heterogeneous Catalytic Domino Reaction

Shainaz M. Landge and Béla Török

Most organic compounds used as active ingredients in drugs are synthesized by the conventional approach of organic chemists – multistep synthesis. With this approach we complete a single chemical transformation in each step, and the preparation of even difficult compounds can eventually be completed by a carefully designed synthetic route. Unfortunately, sometimes it means 10 to 15 or even more reaction steps. However, it also means that every intermediate product has to be isolated and sometimes purified, and all of the additional steps such as isolation, recrystallization, purification and so on significantly add to the accumulation of waste, such as by-products and solvents, which in the best case can be recycled. One could minimize the waste formation and energy consumption if more than one reaction were combined and carried out in one reaction vessel. When these reactions follow a particular sequence we call them domino reactions because of their similarity to the sequence of falling dominos in a row. According to Tietze, in a domino reaction more than one bond will form (usually C–C bonds) under the same reaction conditions without adding reagents or catalysts during the reaction, and the sequence of subsequent reactions is determined by the functionality formed in the previous step [1]. This is in contrast to another type of multiple reactions, namely tandem reactions, where the sequence of steps is not determined and certain steps can occur in a different order. Such reactions are quite common in nature, usually in the form of biosynthetic pathways. In the laboratory the design of such reactions is possible by carefully selecting the functional groups and reaction conditions. One way is to use bifunctional catalysts that are able to catalyze a sequence of reactions. To follow the principles of green chemistry, such catalysts are preferably of a heterogeneous nature. This way the approach would incorporate the inherent advantages of heterogeneous catalysis (as described in Experiment 7) with the efficacy of domino reactions. Probably the most common examples are the applications of metal–solid acid bifunctional catalysts [2]. These catalysts can be made by depositing a metal onto solid acid surfaces, or simply by a thorough mixing of a supported metal catalyst with a solid acid. We have recently developed such a catalyst, which was able to catalyze multistep domino reactions in one reaction vessel [3].

Pyrazoles, an important group among heteroaromatics, are particularly known for their wide range of biological activity [4]. They have been commercially used as pharmaceuticals, agrochemicals, and dyestuffs, e.g. difenzoquat, a herbicide, or phenylbutazone, an anti-inflammatory agent (Scheme 1).

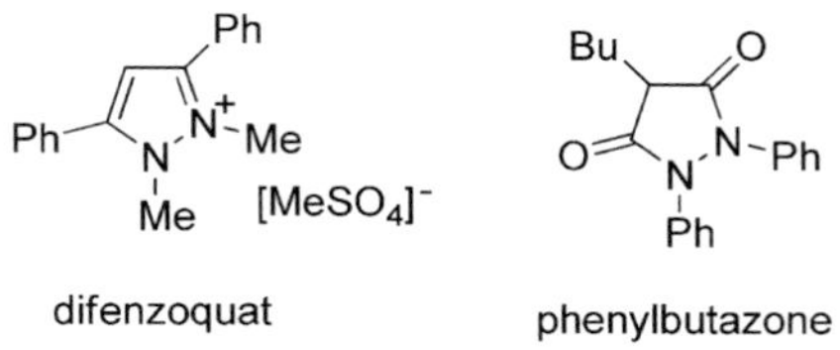

difenzoquat   phenylbutazone

Scheme 1

Because of their importance, a large number of methods are now available for the synthesis of these heteroaromatics. Many of these methods involve multiple-step syntheses as described above and a stoichiometric amount of oxidant to ensure the formation of the heteroaromatic system [5]. However, the use of such reagents is not recommended in view of the tightening of environmental regulations.

Here, a process that combines the advantages of domino reactions, heterogeneous catalysis and microwave irradiation is used for the synthesis of 1,3,5-triphenylpyrazole.

Synthesis of 1,3,5-triphenylpyrazole by a heterogeneous catalytic domino reaction

Preparation of Pd/C/K-10

Apparatus	5 mL glass vial with a cap, stir bar, magnetic stirrer, safety glasses, laboratory coat, protective gloves.
Chemicals	Pd/C (10% palladium on carbon) and K-10 montmorillonite clay.

Attention!   Safety glasses and protective gloves must be worn at all times.

Caution!   Both Pd/C and K-10 are finely powdered materials. Inhalation of the powders can cause irritation to the respiratory tract. Contact with the skin could also cause irritation. All reactions should be carried out in a well-ventilated hood.

Procedure   In a 10 mL glass vial, mix K-10 montmorillonite (500 mg) with Pd/C (21 mg). Stir the dry mixture well on a magnetic stirrer for 5 min to ensure thorough mixing.

Synthesis of 1,3,5-triphenylpyrazole

Apparatus   50 mL round-bottomed flask with stir bar, magnetic stirrer, microwave vial with a cap, syringe (500 µL), CEM Discover Benchmate microwave reactor, safety glasses, laboratory coat, protective gloves.

Note: The reaction is carried out in a top-loader one-vessel Discover Benchmate instrument (CEM). The temperature is monitored by an IR controller. It is anticipated that the experiments can be carried out in oven style instruments such as the MARS (CEM) or START (Milestone) systems, which are equipped with fiberoptic temperature controllers.

Chemicals   Chalcone, phenylhydrazine, dichloromethane. Pd/C/K-10 catalyst.

Attention!   Safety glasses and protective gloves must be worn at all times.

Caution!   Because of the toxicity and volatility of the solvent and the reactants, care should be taken to avoid inhalation of dichloromethane, chalcone, phenylhydrazine or the catalyst prepared above or contact of their solution with the skin. All reactions should be carried out in a well-ventilated hood.

Procedure   The mixture of chalcone (208.2 mg, 1.0 mmol) and phenylhydrazine (148 µL, 1.5 mmol) is dissolved in dichloromethane (3 mL) in a dry 50 mL round-bottomed flask. Then the mechanically premixed combination of 21 mg Pd/C and K-10 (500 mg) (see above) is added. After 10 min stirring, the solvent is evaporated *in vacuo*. The dry mixture is placed in a reaction vial and irradiated in the microwave reactor (CEM Discover Benchmate) at 160 °C for 30 min with stirring. When the reaction is complete, dichloromethane is added to the cold mixture, the suspension is stirred for 10 min, and the catalyst is removed by filtration. The product is purified by recrystallization or flash chromatography (hexane:dichloromethane 80:20) as desired and as time allows. Colorless crystals of 1,3,5-triphenylpyrazole (281 mg, 95 %) can be isolated (Scheme 2).

Characterization data: Colorless crystals, Yield: 95%. m.p. 137–139 °C (hexane:dichloromethane 80:20). ^{1}H NMR (300.1 MHz, CDCl$_3$): δ 7.93 (dd, 2H, J = 7.2, 1.5 Hz, Ar), 7.27-7.45 (m, 13H, Ar), 6.82 (s, 1H, CH) ppm. ^{13}C NMR (75.5 MHz, CDCl$_3$): δ 151.9, 144.3, 140.1, 133.0, 130.5, 128.9, 128.7, 128.6, 128.4, 128.2, 127.9, 127.4, 125.8, 125.2, 105.2 ppm. IR (neat, cm^{-1}): $\bar{v}$ 3119, 3060, 1624, 1596, 1546, 1496, 1482, 1456, 1434, 1415, 1362, 1311, 1284, 1214, 1173, 1066, 1020. MS: m/z 296 (M$^+$, 100%).

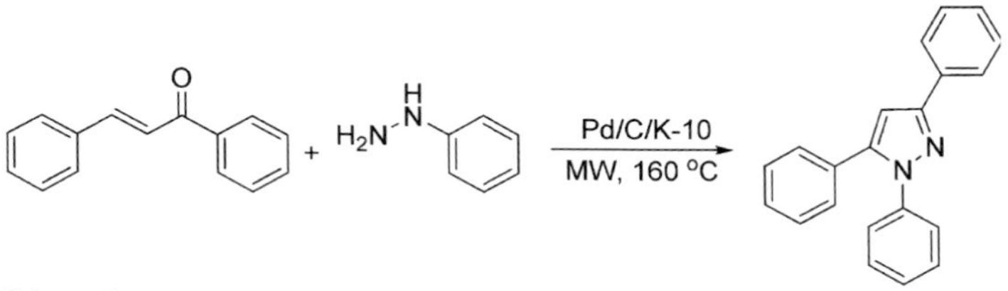

Scheme 2

Waste Disposal Put the recovered dichloromethane and hexane–dichloromethane mixture in a recovery container labeled for halogenated solvents. Collect the catalyst from the filter and put it in a designated container.

Explanation

The basic concept used in this synthesis is to combine the coupling (condensation), the subsequent cyclization and the final oxidation reaction into a three steps/one pot approach by using a special bifunctional noble metal/solid acid catalyst. The result indicates that the concept of designing a one-pot approach is feasible.

The proposed model for the reaction mechanism involves contribution from both the solid acid and the metal. The first step is the activation of chalcone by K-10 montmorillonite (depicted as δ^+ – a Lewis acid center). This material is a widely used catalyst in environmentally friendly synthetic processes (Scheme 3) [6].

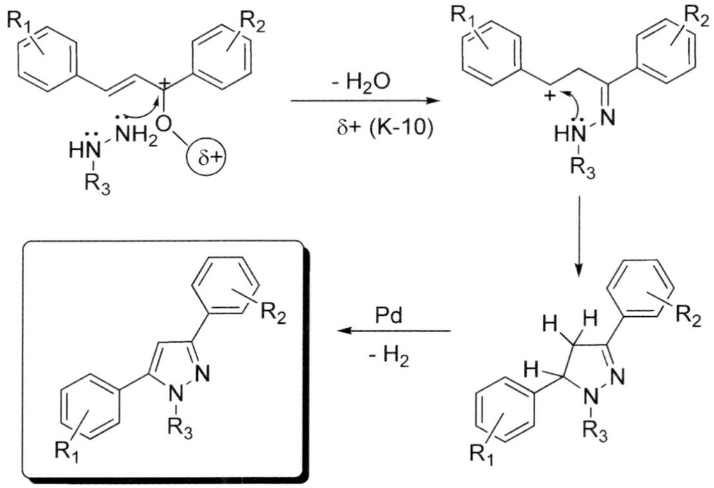

Scheme 3

The use of this material rather than liquid acids is highly preferable; it is solid, noncorrosive, inexpensive and recyclable. In addition, it is commercially available and can be used without any pretreatment. It is an excellent catalyst for microwave-assisted organic synthesis (MAOS), an area that has attracted significant interest in recent years [7]. The chalcone activation occurs at the carbonyl group, which results in the formation of a carbocationic intermediate and continues as a traditional condensation reaction with the NH_2 group of the phenylhydrazine. The catalyst then produces a benzylic cation and the hydrazone undergoes a cyclization reaction to form dihydropyrazole. In the last step the metal component takes a leading role and dehydrogenates the intermediate to the final aromatic product.

References

1 L.F. Tietze, *Chem. Rev.* **1996**, *96*, 115.
2 (a) G. Szöllösi, B. Török, L. Baranyi, M. Bartók, *J. Catal.* **1998**, *179*, 619; (b) B. Török, G. London, M. Bartók, *Synlett* **2000**, 631.
3 S.M. Landge, A. Schmidt, V. Outerbridge, B. Török, *Synlett* **2007**, 1600.
4 (a) L. Xiao, X.Y. Lu, D.M. Ruden, *Mini-Rev. Med. Chem.* **2006**, *6*, 1137; (b) E. McDonald, K. Jones, P.A. Brough, M.J. Drysdale, P. Workman, *Curr. Top. Med. Chem.* **2006**, *6*, 1193.
5 (a) J. Elguero, In *Comprehensive Heterocyclic Chemistry II*, A.R. Katritzky, C.W. Rees, E.F.V. Scriven, Eds.; Pergamon Press, Oxford, **1996**; Vol. 3, p 1; (b) T.L. Gilchrist, *Heterocyclic Chemistry*, 3rd ed. Addison, Wesley, Longman, Edinburgh, **1997**.
6 S. Dasgupta, B. Török, *Org. Prep. Proced. Int.* **2008**, *40*, 1.
7 (a) A. Loupy, *Microwaves in Organic Synthesis*, 2nd ed., Wiley-VCH, **2006**; (b) C.O. Kappe, A. Stadler, *Microwaves in Organic and Medicinal Chemistry*, Wiley-VCH, **2005**.

9

Chemoselective Synthesis of Acylals (1,1-Diesters) from Aldehydes: a Discovery-Oriented Green Organic Chemistry Laboratory Experiment

Michael P. DeVore, Matthew G. Huddle and Ram S. Mohan

Acylals (geminal diacetates) have often been used as protecting groups for carbonyl compounds, because they are stable to neutral and basic conditions [1, 2]. Hence, methods for their synthesis have received considerable attention. However, most introductory organic chemistry texts and laboratory curricula do not introduce students to acylals. In addition, although green chemistry principles are increasingly stressed in the undergraduate curriculum, there are only a few laboratory experiments [3] wherein the toxicity of reagents is taken into consideration in the design of the experiment. Herein we report a discovery-oriented green chemistry experiment that illustrates the selective formation of acylals from aldehydes in the presence of ketones. This experiment illustrates several Green Chemistry principles [4] such as the use of a relatively nontoxic catalyst, solvent-free reaction conditions, and a single-step reaction. The added element of discovery is likely to increase student enthusiasm. The experiment also introduces students to the concept of chemoselectivity.

Recently, bismuth compounds have attracted attention because of their remarkably low toxicity, low cost, and relative insensitivity to air and small amounts of moisture [5, 6]. We have previously reported the bismuth(III) trifluoromethanesulfonate $(Bi(SO_3CF_3)_3)$ catalyzed formation of acylals from aldehydes (Scheme 1) [7, 8].

Based on this methodology, we have developed a discovery-oriented green organic chemistry laboratory experiment in which a mixture of benzaldehyde and acetophenone is reacted with acetic anhydride in the presence of bismuth(III) triflate, $Bi(SO_3CF_3)_3$, under solvent-free conditions. Under these reaction condi-

Scheme 1 $OTf = SO_3CF_3$.

tions, the aldehyde is converted into the corresponding acylal while the ketone remains unchanged. The product mixture is analyzed by ^{1}H and ^{13}C NMR spectroscopy. Using a series of lead questions as cues, students are asked to figure out the identity of the product mixture. The possible outcomes of the reaction and the actual product formed are shown in Scheme 2. The short reaction time and ease of workup allow the experiment to be completed in 1 h.

Scheme 2 OTf = SO$_3$CF$_3$.

Acylal Preparation

Apparatus   Large vial, separatory funnel, recovery flask, rotary evaporator, safety glasses, laboratory coat, protective gloves.

Chemicals   Benzaldehyde, acetophenone, acetic anhydride, bismuth triflate, diethyl ether, Na$_2$CO$_3$, NaCl, Na$_2$SO$_4$, CDCl$_3$. (All reagents were purchased from Aldrich Chemical Company except bismuth triflate, which was obtained from Lancaster Chemicals.)

Attention!   Safety glasses and protective gloves must be worn at all times.

Caution!   Diethyl ether vapor is harmful and very flammable. All reactions should be carried out in a well-ventilated hood.

Procedure   In a large vial, a mixture of benzaldehyde (0.1020 g, 0.961 mmol), acetophenone (0.1106 g, 0.921 mmol) and acetic anhydride (0.353 g, 0.327 mL, 3.46 mmol, 3.6 equivalents) is stirred and cooled in an ice bath as bismuth triflate (0.0128 g, 0.0192 mmol, 2.0 mol%) is added. The resulting yellow mixture is stirred for 30 min. Then 10% aqueous Na$_2$CO$_3$ (5.0 mL) solution is added and the mixture is stirred for 10 min. The reaction mixture is then transferred to a separatory funnel and extracted with diethyl ether (2 × 10 mL). The combined organic layers are washed with saturated aqueous NaCl (15.0 mL) solution, dried (Na$_2$SO$_4$) and concentrated on a rotary evaporator to yield 0.145 g (72%) of a yellow liquid that is analyzed by ^{1}H and ^{13}C NMR spectroscopy.

Characterization data: Benzaldehyde: ^{1}H NMR (270 MHz, CDCl$_3$): δ 10.0 (s, 1H), 7.87 (m, 2H), 7.57 (m, 3H) ppm. ^{13}C NMR (67.5 Hz): δ 192.2, 136.2, 134.2, 129.5, 128.8 ppm.

Acetophenone: ^{1}H NMR (CDCl$_3$): δ 7.95 (m, 2H), 7.49 (m, 3H), 2.59 (s, 3H) ppm. ^{13}C NMR (67.5 Hz): δ 197.6, 136.6, 132.7, 128.1, 127.9, 26.1 ppm.

Benzaldehyde acylal: ^{1}H NMR (70 MHz, CDCl$_3$): δ 7.67 (s, 1H, CH(OAc)$_2$), 7.51 (m, 2H), 7.39 (m, 2H), 2.11 (s, 3H) ppm. ^{13}C NMR (67.5 Hz): δ 168.5, 135.3, 129.5, 128.4, 126.4, 89.4, 20.5 ppm.

Waste Disposal The aqueous layer must be placed in an aqueous waste jar. The unused product can be dissolved in a minimal amount of acetone and placed in a nonhalogenated waste container. The NMR solution should be disposed of in a halogenated waste container.

Questions

Students are asked to answer the following questions pertaining to the NMR spectra of the product:

1. Is there a resonance for an aldehydic hydrogen in the ^{1}H NMR spectrum? Is there a resonance for an aldehydic carbon in the ^{13}C NMR spectrum?

   The aldehydic hydrogen resonance (^{1}H NMR) and aldehydic carbon resonance (^{13}C NMR) are no longer present in the product, indicating that benzaldehyde has reacted.

2. Is there a resonance for a ketone methyl group in the ^{1}H NMR spectrum? Is there a resonance for a ketone carbon in the ^{13}C NMR spectrum?

   The ketone methyl group resonance and the carbonyl resonance are still present indicating that acetophenone has not reacted.

3. Are there any new carbonyl resonance(s) in the ^{13}C NMR spectrum?

   A resonance at δ 168.6 suggests the formation of an ester. A singlet in the ^{1}H NMR at δ 2 suggests a methyl ester.

Notes to the instructor

1. The discovery element in this experiment can be presented in two ways: (a) Students can be told what the possible products are and then asked to figure out which compound (benzaldehyde or acetophenone) reacts *or* (b) they can be asked to determine what the product is and which compound gives rise to it.
2. If a rotary evaporator is not available, the solvent can be removed in a hot water bath.

References

1 T.W. Greene, P.G.M. Wuts, *Protective Groups in Organic Synthesis,* 3rd ed.; John Wiley and Sons, Inc: New York, **1999**.

2 M.J. Gregory, *J. Chem. Soc. B* **1970**, 1201.

3 For some recent articles and laboratory experiments pertaining to green chemistry, see (a) J.E. Christensen, M.G. Huddle, J.L. Rogers, H. Yung, *J. Chem. Educ.* **2008**, *85*, 1274; (b) G.D. Bennett, *J. Chem. Educ.* **2006**, *83*, 1871; (c) A.J.F.N. Sobral, *J. Chem. Educ.* **2006**, *83*, 1665; (d) J. Pereira, C.A.M. Afonso, *J. Chem. Educ.* **2006**, *83*, 1333; (e) B. Braun, R. Charney, A. Clarens, J. Farrugia, C. Kitchens, C. Lisowski, D. Naistat, A. O'Neil, *J. Chem. Educ.* **2006**, *83*, 1126; (f) J. Bennett, K. Meldi, C. Kimmell, II, *J. Chem. Educ.* **2006**, *83*, 1221; (g) K.K.W. Mak, J. Siu, Y.M. Lai, P. Chan, *J. Chem. Educ.* **2006**, *83*, 943; (h) I. Montes, D. Sanabria, M. García, J. Castro, J. Fajardo, *J. Chem. Educ.* **2006**, *83*, 628; (i) M.R. Dintzner, P.R. Wucka, T.W. Lyons, *J. Chem. Edu.* **2006**, *83*, 270.

4 P.T. Anastas, J.C. Warner, *Green Chemistry: Theory and Practice,* Oxford University Press: New York, **1998**.

5 Most bismuth(III) compounds have an LD_{50} value that is comparable to or even less than that of NaCl.

6 For reviews see (a) J.A. Marshall, *Chemtracts* **1997**, 1064; (b) H. Suzuki, T. Ikegami, Y. Matano, *Synthesis* **1997**, 249; (c) Organobismuth Chemistry; H. Suzuki, Y. Matano. Ed.; Elsevier: Amsterdam, 2001; (d) N.M. Leonard, L.C. Wieland, R.S. Mohan, *Tetrahedron* **2002**, *58*, 8373; (e) H. Gaspard-Iloughmane, C. Le Roux, *Eur. J. Org. Chem.* **2004**, 2517.

7 M.D. Carrigan, K.J. Eash, M.C. Oswald, R.S. Mohan, *Tetrahedron Lett.* **2001**, *42*, 8133.

8 For some recent examples of synthesis of acylals using benign reagents/conditions see (a) V.T. Kamble, V.S. Jamode, N.S. Joshi, A.V. Biradar, R.Y. Deshmukh, *Tetrahedron Lett.* **2006**, *47*, 5573; (b) V.T. Kamble, R.A. Tayade, B.S. Davane, K.R. Kadam, *Austr. J. Chem.* **2007**, *60*, 590; (c) M.A.F.M. Rahman, Y. Jahng, *Eur. J. Org. Chem.* **2007**, 379; (d) J.R. Satam, R.V. Jayaram, *Catal. Commun.* **2007**, *8*, 1414; (e) J. Wang, L. Yan, G. Qian, K. Yang, H. Liu, X. Wang, *Tetrahedron Lett.* **2006**, *47*, 8309.

10

The Effect of a Catalyst on the Reaction of *p*-Hydroxybenzaldehyde with Acetic Anhydride: a Discovery-Oriented Green Organic Chemistry Laboratory Experiment

Matthew G. Huddle, Michael P. DeVore and Ram S. Mohan

Phenols readily react at room temperature with acetic anhydride to yield the corresponding acetates. However, the reaction of aldehydes with acetic anhydride at room temperature to yield the corresponding acylal requires the use of a catalyst. We have developed a laboratory experiment based on the bismuth(III) triflate ($Bi(SO_3CF_3)_3$) catalyzed synthesis of an acylal from an aldehyde in the presence of a ketone [1]. Here we describe a discovery-oriented green chemistry laboratory experiment that illustrates the effect of a catalyst, bismuth triflate, on the reaction of *p*-hydroxybenzaldehyde with acetic anhydride (Schemes 1 and 2). Both reactions illustrate green principles [2] such as solvent-free reaction conditions and the use of no catalyst or a relatively nontoxic catalyst.

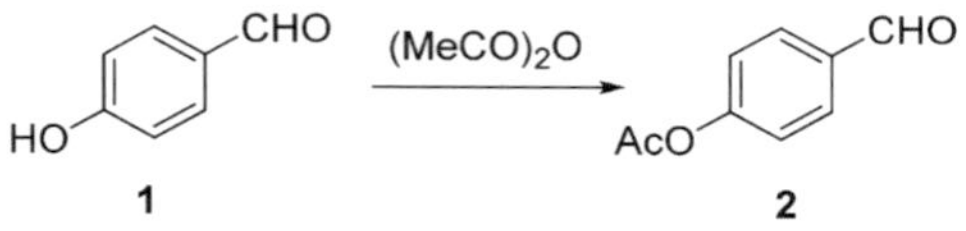

Scheme 1

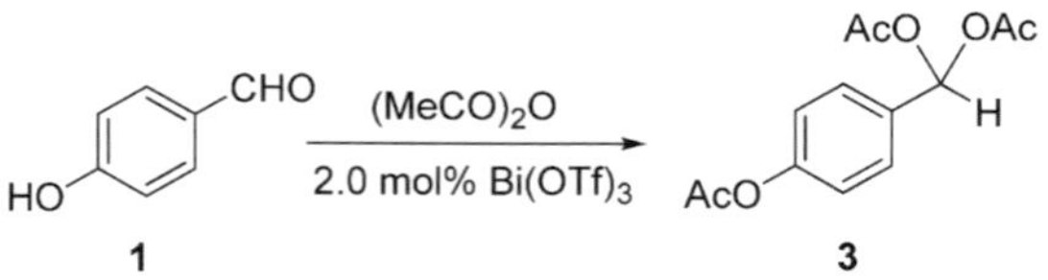

Scheme 2   OTf = SO_3CF_3.

When *p*-hydroxybenzaldehyde is reacted with acetic anhydride in the absence of the catalyst, the product is *p*-acetoxybenzaldehyde. In the presence of bismuth(III) triflate (2.0 mol%), both the phenol and aldehyde undergo reaction to produce **3**. The products are analyzed by ^1H and ^{13}C NMR spectroscopy. Using a series of lead questions as cues, students are asked to figure out the identity of the product mixture.

Reaction of *p*-hydroxybenzaldehyde with acetic anhydride

Apparatus   Large vial, separatory funnel, recovery flask, rotary evaporator, safety glasses, laboratory coat, protective gloves.

Chemicals   *p*-Hydroxyenzaldehyde, acetic anhydride, bismuth triflate, diethyl ether, Na_2CO_3, NaCl, Na_2SO_4, $CDCl_3$. (All reagents were purchased from Aldrich Chemical Company except bismuth triflate, which was obtained from Lancaster Chemicals.)

Attention!   Safety glasses and protective gloves must be worn at all times.

Caution!   Diethyl ether vapor is harmful and very flammable. All reactions should be carried out in a well-ventilated hood.

Procedure   *Reaction of p-hydroxybenzaldehyde with acetic anhydride in the absence of catalyst*

In a vial, a mixture of *p*-hydroxybenzaldehyde (0.2176 g, 1.782 mmol) and acetic anhydride (0.546 g, 0.51 mL, 5.346 mmol, 3.0 equivalents) is stirred at room temperature. After 45 min 10% aqueous Na_2CO_3 (5.0 mL) solution is added and the mixture is stirred for 10 min. The reaction mixture is then transferred to a separatory funnel and extracted with diethyl ether (2×10 mL). The combined organic layers are washed with saturated aqueous NaCl (15.0 mL), dried (Na_2SO_4) and concentrated on a rotary evaporator to give 0.172 g (59% yield) of a clear yellow liquid that is analyzed by ^1H and ^{13}C NMR spectroscopy.

Characterization data: ^1H NMR (270 MHz, $CDCl_3$) *p*-acetoxybenzaldehyde **2**: δ 9.88 (s, 1H), 7.82 (m, 2H), 7.19 (m, 2H), 2.24 (s, 3H) ppm. ^{13}C NMR (67.5 Hz): δ 190.8, 168.5, 155.1, 133.7, 130.9, 122.1, 20.9 ppm.

Waste Disposal The aqueous layer must be placed in an aqueous waste jar. The unused product can be dissolved in a minimal amount of acetone and placed in a nonhalogenated waste container. The NMR solution should be disposed of in a halogenated waste container.

Reaction of p-hydroxybenzaldehyde with acetic anhydride in the presence of Bi(SO₃CF₃)₃

Reaction of p-hydroxybenzaldehyde with acetic anhydride in the presence of $Bi(SO_3CF_3)_3$

In a vial, a mixture of p-hydroxybenzaldehyde (0.1533 g, 1.255 mmol) and acetic anhydride (0.385 g, 0.36 mL, 3.77 mmol, 3.0 equivalents) is stirred at room temperature as $Bi(SO_3CF_3)_3$ (0.0165 g, 0.252 mmol, 2.0 mol%) is added. After 45 min, 10% aqueous Na_2CO_3 (5.0 mL) solution is added and the mixture is stirred for 10 min. The reaction mixture is then transferred to a separatory funnel and extracted with diethyl ether (2 × 10 mL). The combined organic layers are washed with saturated aqueous NaCl (15.0 mL), dried (Na_2SO_4) and concentrated on a rotary evaporator to give 0.253 g (76% yield) of a white solid that is analyzed by ^1H and ^{13}C NMR spectroscopy.

Characterization data: ^1H NMR (270 MHz, CDCl₃) p-acetoxybenzaldehyde acylal **3**: δ 7.61 (s, 1H), 7.48 (m, 2H), 7.06 (m, 2H), 2.21 (s, 3H), 2.03 (s, 6H) ppm. ^{13}C NMR (67.5 Hz): δ 168.8, 168.4, 151.3, 132.8, 127.7, 121.5, 88.8, 20.7, 20.4 ppm.

Waste Disposal The aqueous layer must be placed in an aqueous waste jar. The unused product can be dissolved in a minimal amount of acetone and placed in a nonhalogenated waste container. The NMR solution should be disposed in a halogenated waste container.

Notes to the instructor

1. The discovery element in this experiment can be presented in two ways: (a) Students can be told what the possible products are and then asked to figure out which group (phenol or aldehyde) react(s) or (b) they can be asked to determine what the product is.
2. If a rotary evaporator is not available, the solvent can be removed in a hot water bath.

Reference

1 H.W. Roesky, D.K. Kennepohl, *Experiments in Sustainable Chemistry*, Wiley-VCH, Weinheim, **2009**, Ch. 9.

2 P.T. Anastas, J.C. Warner, *Green Chemistry: Theory and Practice*, Oxford University Press: New York, **1998**.

11

Renewable Chemicals by Sustainable Oxidations Using Gold Catalysts

Søren K. Klitgaard, Yury Gorbanov, Esben Taarning and Claus H. Christensen

The oxidation of alcohols (or aldehydes) is frequently used in the chemical industry, as well as in traditional organic synthesis, to produce the corresponding aldehydes, ketones and carboxylic acid derivatives [1] . Many of these oxidations are usually carried out with stoichiometric amounts of high-valent metal oxides, e.g. manganese or chromium oxides, which produce a huge amount of waste in terms of metal ions. To avoid this, considerable effort has been invested to develop aerobic oxidations as a possible alternative. In aerobic oxidations, air or pure dioxygen are used as the oxidants instead of the classical metal oxides. Dioxygen is considered a »green« oxidant because it produces water as the only by-product. From an economic point of view, dioxygen is also very attractive because of its low cost and unlimited accessibility. However, to achieve selective oxidations with dioxygen at reasonably low temperatures, a suitable catalyst is needed.

Metallic gold was considered to be useless in heterogeneous catalysis for many years because of its inertness. However, this changed with the discovery that gold nanoparticles can indeed be very active in oxidation reactions. This was first reported by Haruta et al. for the low-temperature aerobic oxidation of carbon monoxide [2]. Following the discovery by Haruta et al., a variety of other aerobic oxidation reactions were discovered. For instance, gold nanoparticles were used in the oxidation of alcohols to aldehydes, carboxylic acids or esters, in the oxidation of aldehydes to esters, in epoxidations of olefins, and in the oxidation of amines to amides [3–18]. Here, we illustrate how an active and selective aerobic oxidation catalyst can be conveniently produced, and how it can be used to prepare important chemical products from renewable raw materials.

Sustainable Oxidations Using Gold Catalysts

Preparation of K$_2$Ti$_6$O$_{13}$ support

Apparatus 220 mL stainless steel Teflon-lined autoclave, magnet, magnetic stirrer, oven, safety glasses, laboratory coat, protective gloves.

Chemicals TiO$_2$ anatase nanoparticles, KOH (10 mol L^{-1}).

Attention! Safety glasses and protective gloves must be worn at all times.

Caution! Concentrated KOH is very corrosive and may damage skin and eyes with contact. Because of explosion danger, care should be taken when heating autoclaves.

Procedure In a 220 mL stainless steel Teflon-lined autoclave, TiO$_2$ (1.92 g) is suspended in the concentrated aqueous KOH solution (10 mol L^{-1}, 160 mL) with stirring. The autoclave is closed and placed in an oven at 150 °C for 3 days. During the hydrothermal treatment, the TiO$_2$ nanoparticles are transformed into K$_2$Ti$_6$O$_{13}$ nanowires. The resulting suspension is filtered and the powder is washed with large amounts of distilled water until neutral pH is achieved in the washing water. The K$_2$Ti$_6$O$_{13}$ nanowires are dried in air overnight.

Characterization data: BET surface area ~309 m^2 g^{-1}. XRPD: phase-pure K$_2$Ti$_6$O$_{13}$.

Preparation of 1 wt% Au on K$_2$Ti$_6$O$_{13}$ catalyst

Apparatus 50 mL beaker, magnetic stirring bar, magnetic stirrer, oven, safety glasses, laboratory coat, protective gloves.

Chemicals H[AuCl$_4$]•3H$_2$O, water, saturated solution of NaHCO$_3$, support K$_2$Ti$_6$O$_{13}$ (TiO$_2$ anatase nanoparticles may be used instead). H$_2$/N$_2$ gas mixture (any mixture of dihydrogen, preferably more than 5% H$_2$, in an inert gas can be used.

Attention! Safety glasses and protective gloves must be worn at all times.

Procedure In a 50 mL beaker, H[AuCl$_4$] •3H$_2$O (0.10 g) is dissolved in water (30 mL) resulting in a pale yellow solution. A saturated solution of NaHCO$_3$ (ca. 30 drops) is added until pH = 9 under stirring. Immediately after this, the support (4.95 g) is added to the solution. The suspension is stirred at 50 °C for approximately 1 h. During this time the aqueous solution gradually shifts from yellow to clear as Au$_2$O$_3$•xH$_2$O precipitates onto the support. After 1 h, the suspension is filtered and the catalyst is washed with water until the washing water is free of Cl$^-$ ions. This can be tested by adding a few drops of AgNO$_3$ solution to a portion of the washing water, which leads to precipitation of AgCl if any Cl$^-$ ions are present.

The precipitated gold oxides are reduced using H$_2$ in N$_2$ at 300 °C for 2 h with a heating ramp of 5 °C min^{-1}. The resulting dark purple powder consists of gold nanoparticles with a size of around 4–5 nm.

Tip: It is possible to use TiO$_2$ instead of K$_2$Ti$_6$O$_{13}$, but the catalytic activity will be somewhat lower. When using TiO$_2$, the reduction of the gold oxides can also be done by calcination in air at 450 °C for 4 h.

Catalytic aerobic oxidation of benzyl alcohol to produce methyl benzoate

Apparatus 25 mL round-bottomed flask, condenser, magnetic stirring bar, magnetic stirrer, safety glasses, laboratory coat, protective gloves.

Chemicals Benzyl alcohol, methanol, KOH in methanol solution (1 mol L^{-1}), air or O$_2$ cylinder, gold catalyst prepared as outlined above.

Attention Safety glasses and protective gloves must be worn at all times.

Procedure A mixture of benzyl alcohol (0.414 mL, 4 mmol), methanol (4.45 mL, 110 mmol), and KOH in methanol (0.400 mL, 0.40 mmol KOH) is charged to a 25 mL round-bottomed flask equipped with a condenser connected to the O$_2$ cylinder. The system is flushed with O$_2$, and the catalyst (0.098 g) is added. The reaction is carried out at atmospheric pressure in O$_2$. The suspension is stirred for approximately 20 h, and it is possible to follow the reaction by withdrawing samples during this period of time. The catalyst is filtered off and the methyl benzoate is isolated by distillation under vacuum (yield ca. 90%).

Tip: The amounts of substrates and products can be quantified with GC using an internal standard like anisole or decane. It is possible to use air as the oxidant instead of pure dioxygen, but the yield could be slightly lower.

Characterization data: b.p. 199.6 °C. MS: *m/z* 136 (M⁺, 35%), 105 (C₆H₅CO, 100%).

Catalytic aerobic oxidation of benzaldehyde to produce methyl benzoate

Apparatus   25 mL round-bottomed flask, condenser, magnet, magnetic stirrer, safety glasses, laboratory coat, protective gloves.

Chemicals   Benzaldehyde, methanol, KOH in methanol solution (1 mol L⁻¹), air or O₂ cylinder, gold catalyst prepared as outlined above.

Attention   Safety glasses and protective gloves must be worn at all times.

Procedure   A mixture of benzaldehyde (0.408 mL, 4 mmol), methanol (4.45 mL, 110 mmol) and KOH in methanol (0.400 mL, 0.40 mmol KOH) is charged to a 25 mL round-bottomed flask equipped with a condenser connected to an O₂ cylinder. The system is flushed with O₂ and the catalyst (0.098 g) is added. The reaction is carried out at atmospheric pressure in O₂. The suspension is stirred for approximately 1 h, and it is possible to follow the reaction by withdrawing samples during this period of time. The catalyst is filtered off and the methyl benzoate is isolated by distillation under vacuum (Yield ca. 92%).

Tip: The amounts of substrates and products can be quantified with GC using an internal standard like anisole or decane. It is possible to use air as the oxidant instead of pure dioxygen, but the yield could be lower.

Characterization data: b.p. 199.6 °C. MS: *m/z* 136 (M⁺, 35%), 105 (C₆H₅CO, 100%).

Catalytic aerobic oxidation of 5-hydroxymethyl methylfuroate

Apparatus   25 mL round-bottomed flask, condenser, magnet, magnetic stirrer, safety glasses, laboratory coat, protective gloves.

Chemicals	5-(Hydroxymethyl)furfural, methanol, 25 wt% sodium methoxide solution, air or O_2 cylinder, gold catalyst prepared as outlined above.

Attention Safety glasses and protective gloves must be worn at all times.

Procedure A mixture of 5-(hydroxymethyl)furfural (0.504 g, 4 mmol), methanol (4.45 mL, 110 mmol), and a solution of sodium methoxide in methanol (0.069 mL, 0.30 mmol base) is charged into a 25 mL round-bottomed flask equipped with a condenser connected to an O_2 cylinder. The system is flushed with O_2 and the catalyst (0.25 g) is added. The reaction is carried out at atmospheric pressure in O_2. The suspension is stirred for approximately 20 h, and it is possible to follow the reaction by withdrawing samples during this period of time. The catalyst is filtered off and the 5-hydroxymethyl methylfuroate is isolated by distillation under vacuum (yield ca. 89%).

Tip: The amounts of substrates and products can be quantified with GC using an internal standard like anisole or decane. It is possible to use air as oxidant instead of pure oxygen, but the yield could be slightly lower.

Characterization data: MS: m/z 156 (M$^+$, 38 %), 97 ($C_4H_2OCH_2OH$, 100 %).

Catalytic aerobic oxidation to prepare furan-2,5-dimethylcarboxylate

Apparatus Magnetic stirring bar (or a mechanically stirred autoclave), 325 mL titanium-stabilized T316 autoclave from the Parr Instrument Company, oil bath, magnetic stirrer, magnetic stirring bar, safety glasses, laboratory coat, protective gloves.

Chemicals 5-(Hydroxymethyl)furfural, methanol, 25 wt% sodium methoxide in methanol, O_2 cylinder, gold catalyst prepared as outlined above.

Attention Safety glasses and protective gloves must be worn at all times. The autoclave *must* be equipped with a safety valve. Mixing organic chemicals with dioxygen at high pressures requires the utmost care. The reaction has to be set up in an autoclave room!

Procedure A mixture of 5-(hydroxymethyl)furfural (0.504 g, 4 mmol), methanol (12.65 mL, 300 mmol), gold catalyst (0.25 g), and a solution of sodium methoxide in methanol (0.069 mL, 0.3 mmol base) is charged to a 325 mL titanium-stabilized T316 autoclave from the Parr Instrument Company. The autoclave is fitted with a magnetic stirring bar, flushed with dioxygen, and pressurized with 4 bar O_2 (52 mmol). The autoclave is then immersed in an oil bath kept at 130 °C and left, with stirring, for 6 h. After reaction, the autoclave is cooled to room temperature and the reaction mixture is analyzed using GC/GC-MS after dissolving all organic material in an excess of methanol. The catalyst is filtered off, and the furan-2,5-dimethylcarboxylate is isolated by distilling off the methanol at atmospheric pressure. The remaining solid can be sublimed at 1 atm and 170 °C to afford colorless crystals (yield ca. 81%).

Characterization data: MS: m/z 184 (M^+, 52 %), 153 ($MeOOCC_4H_2OCO$, 100 %). ^1H NMR (300 MHz): δ 3.92 (6H), 7.21 ppm (2H) ppm. ^{13}C NMR: δ 52.39, 118.46, 146.57, 158.37 ppm.

Explanation

It is much easier to separate the excess of the oxidant from the reaction mixture when aerobic oxidations are used than when the oxidations rely on classical metal oxide oxidants, and clearly the amount of waste is much lower. Additionally, the heterogeneous catalyst can be easily removed by filtration and reused.

In the present syntheses, methyl esters result because the reactions are performed in an excess of methanol. This makes it possible to minimize the amount of base added to the reaction mixture. Without base, the reactions are very slow and do not proceed to completion, probably because free carboxylic acids poison the gold catalyst. The present methodology relies on the fact that methanol is more difficult to oxidize than the substrate, but about 0.5–3 moles of methanol can still be lost as CO_2 for each mole of substrate converted, depending on the exact reaction conditions. An excess of methanol is therefore required, and this also suppresses formation of esters other than the methyl ester, e.g. by transesterification of the methyl ester product with the reactant.

Today, a large fraction of the petrochemical feedstocks is used to produce polymers. In attempts to find alternatives, biorenewable feedstocks are being considered with significant interest. 5-(Hydroxymethyl) furfural is believed to be a promising sustainable platform molecule since it can be prepared directly from glucose or fructose. The oxidized derivatives of 5-(hydroxymethyl) furfural are good candidates for replacing terephthalic acid (benzene-1,4-dicarboxylic acid), which is used as a monomer in polyethylene terephthalate (PET) plastic.

References

1 (a) G. Franz, R.A. Sheldon, *Ullmann's Encyclopedia of Industrial Chemistry*, Wiley-VCH, Weinheim, **2005**; (b) K. Weissermel, H.-J. Arpe, *Industrial Organic Chemistry, 3rd ed.*, VCH Verlagsgesellschaft, Weinheim **1997**.
2 M. Haruta, N. Yamada, T. Kobayashi, S. Iijima, *J. Catal.* **1989**, *115*, 301.
3 D.I. Enache, J.K. Edwards, P. Landon, B. Solsona-Espriu, A.F. Carley, A.A. Herzing, M. Watanabe, C.J. Kiely, D.W. Knight, G.J. Hutchings, *Science* **2006**, *311*, 362.
4 A. Abad, P. Concepción, A. Corma, H. Garcia, *Angew. Chem.* **2005**, *117*, 4134; *Angew. Chem. Int. Ed.* **2005**, *44*, 4066.
5 C.H. Christensen, B. Jørgensen, J. Rass-Hansen, K. Egeblad, R. Madsen, S.K. Klitgaard, S.M. Hansen, M.R. Hansen, H.C. Andersen, A. Riisager, *Angew. Chem.* **2006**, *118*, 4764; *Angew. Chem. Int. Ed.* **2006**, *45*, 4648.
6 (a) T. Hayashi, T. Inagaki, N. Itayama, H. Baba, *Catal. Today* **2006**, *117*, 210; (b) I.S. Nielsen, E. Taarning, K. Egeblad, R. Madsen, C.H. Christensen, *Catal. Lett.* **2007**, *116*, 35.
7 C. Marsden, E. Taarning, D. Hansen, L. Johansen, S.K. Klitgaard, K. Egeblad, C.H. Christensen, *Green Chem.* **2008**, *10*, 168.
8 (a) T. Hayashi, K. Tanaka, M. Haruta, *J. Catal.* **1998**, *178*, 566; (b) M. Haruta, M. Daté, *Appl. Catal. A* **2001**, *222*, 427.
9 S.K. Klitgaard, K. Egeblad, U.V. Mentzel, A.G. Popov, T. Jensen, E. Taarning, I.S. Nielsen, C.H. Christensen, *Green Chem.* **2008**, *10*, 419.
10 G.H. Du, Q. Chen, R.C. Che, Z.Y. Yuan, L.-M. Peng, *Appl. Phys. Lett.* **2001**, *79*, 3702.
11 (a) S. Zhang, Q. Chen, L.-M. Peng, *Phys Rev. B* **2005**, *71*, 014104; (b) Q. Chen, G.H. Du, S. Zhang, L.-M. Peng, *Acta Cryst.* **2002**, *B58*, 587.
12 G.H. Du, Q. Chen, P.D. Han, Y. Yu, L.-M. Peng, *Phys. Rev. B* **2003**, *67*, 035323.
13 (a) B. Zhu, K. Li, Y. Feng, S. Zhang, S. Wu, W. Huang, *Catal. Lett.* **2007**, *118*, 55; (b) J. Jiang, Q. Gao, Z. Chen, *J. Mol. Catal. A: Chem.* **2008**, *280*, 233.
14 L. Miao, Y. Ina, S. Tanemura, T. Jiang, M. Tanemura, K. Kaneko, S. Toh, Y. Mori, *Surf. Science* **2007**, *601*, 2792.
15 M. Haruta, *Catal. Today* **1997**, *36*, 153.
16 (a) P. Fristrup, L.B. Johansen, C.H. Christensen, *Catal. Lett.* **2008**, *120*, 184.; (b) B. Jørgensen, S.E. Christiansen, M.L.D. Thomsen, C.H. Christensen, *J. Catal.* **2007**, *251*, 332.
17 N. Zheng, G.D. Stucky, *Chem. Commun.* **2007**, 3862.
18 H. Miyamura, R. Matsubara, Y. Miyazaki, S. Kobayashi, *Angew. Chem.* **2007**, *119*, 4229; *Angew. Chem. Int. Ed.* **2007**, *46*, 4151.

12

Sustainable Ruthenium-Catalyzed Direct Arylations through C–H Bond Functionalizations

Lutz Ackermann and Rubén Vicente

Transition metal-catalyzed cross-coupling reactions between organometallic reagents and organic electrophiles are useful tools for syntheses of unsymmetrically substituted biaryls, which are indispensable substructures of, *inter alia*, biologically active compounds, natural products, liquid crystals, or organic polymers [1, 2]. The organometallic compounds are often not commercially available and give rise to the formation of undesired by-products, which should be minimized for ecological reasons. Syntheses of these organometallic reagents require additional chemical transformations, during which further by-products are unfortunately obtained (Scheme 1, *path a*).

Direct arylation reactions represent economically attractive and ecologically benign alternatives to traditional cross-coupling methodologies (Scheme 1, *path b*). Here, direct functionalizations of C–H bonds not only lead to minimization of reaction steps, but also significantly reduce the formation of undesired waste [1, 3].

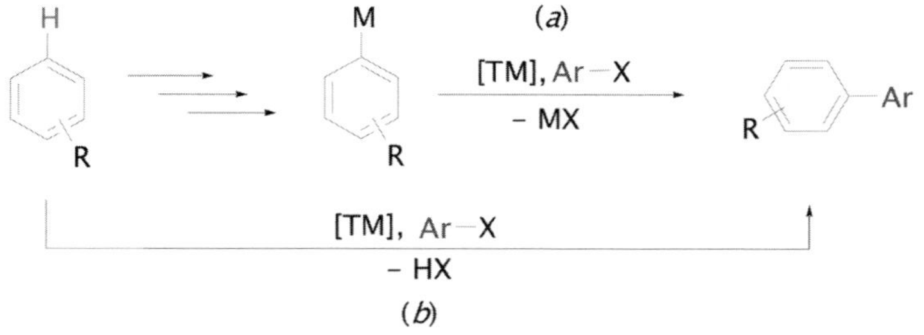

Scheme 1 Sustainable direct arylations as an alternative strategy for biphenyl syntheses.

Generally, arenes as pronucleophilic starting materials display various C–H bonds with comparable bond dissociation energies for direct arylation reactions. Consequently, the development of highly regioselective direct arylations for synthetically useful transformations represents a major challenge.

Regioselectivities of direct arylations can be controlled when the electronic properties of a given pronucleophile dominate its reactivity. This holds true for a variety of heteroarenes, and sets the stage for regioselective palladium-catalyzed C–H bond functionalizations of electron-rich, as well as electron-deficient heteroarenes [1, 3].

In contrast, electronically unbiased arenes often lead to unselective processes which give rise to undesired mixtures of regioisomers which are often difficult to purify. As a result, strategies relying on potentially removable directing groups were developed. Here, Lewis-basic directing groups (DG) coordinate to transition metal catalysts [4], allowing for intramolecular C–H bond cleavages (Scheme 2) [5]. Thereby, regioselective direct arylations can be efficiently accomplished in an overall intermolecular fashion [1, 5].

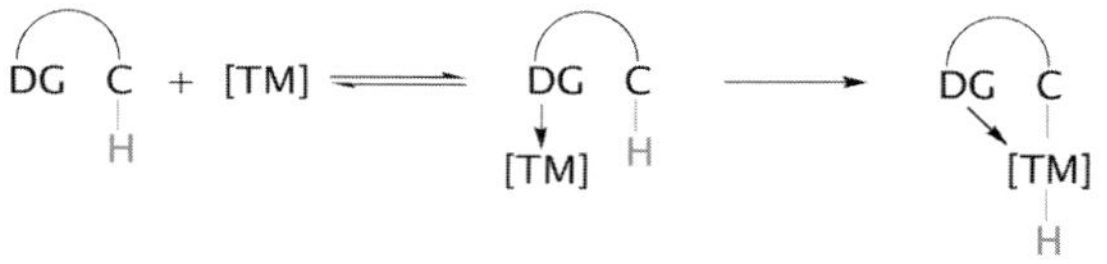

Scheme 2 Regioselective intermolecular C–H bond functionalizations through the use of directing groups (DG).

In recent years, ruthenium-catalyzed chelation-assisted direct arylations have emerged as valuable tools for the regioselective preparation of biphenyl derivatives [5, 6]. Particularly, catalytic systems consisting of [RuCl$_2$(p-cymene)]$_2$ (**1**) and air-stable (heteroatom-substituted) secondary phosphine oxides ((HA)SPOs) [7] as preligands, such as HASPO **2** or SPO (1-Ad)$_2$P(O)H (**3**), enabled broadly applicable C–H bond functionalizations with high catalytic efficacy (Scheme 3). Thereby, a variety of pronucleophilic arenes could be converted with aryl bromides, chlo-

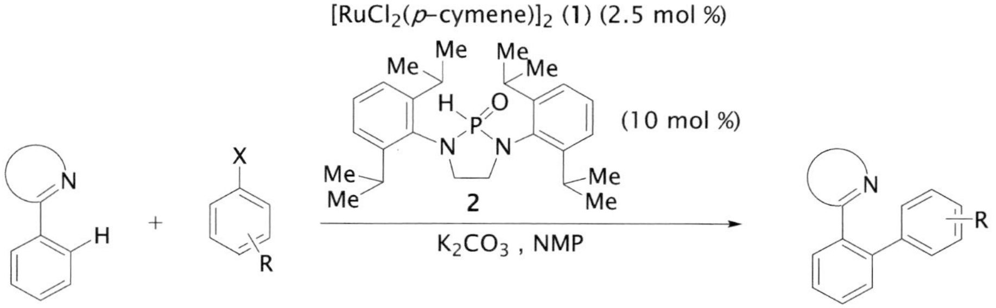

Scheme 3 Ruthenium-catalyzed chelation-assisted direct arylations with HASPO **2** as preligand.

rides or even tosylates, bearing both electron-withdrawing and electron-donating substituents on the arenes of the electrophiles [6b, 7, 8].

Generally, ruthenium-catalyzed direct arylations with organic electrophiles were thus far limited to the use of highly polar N-methyl-2-pyrrolidinone (NMP) as solvent [1, 5]. However, more efficient catalysts have recently been disclosed which enabled these C–H bond functionalizations in apolar solvents such as toluene. Thus, a catalytic system comprising ruthenium precursor 1 and SPO 3 as preligand allowed for a regioselective direct arylation of triazole 4 with aryl bromide 5 in toluene as solvent (Scheme 4) [9].

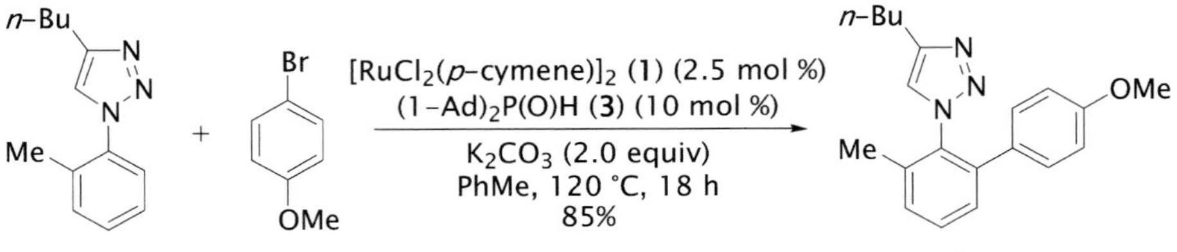

Scheme 4 Ruthenium-catalyzed direct arylation of triazole 4 in toluene with SPO 3 as preligand.

Of all the aryl halides, the chlorides arguably constitute the most useful single class of substrates because of the lower costs associated with their use and the wide diversity of (commercially) available compounds [1]. While sustainable ruthenium-catalyzed direct arylations with aryl chlorides as electrophiles could be accomplished using SPO 3 in NMP, mesitylene carboxylic acid (7) as preligand proved to generate a broadly applicable ruthenium catalyst when using apolar toluene as solvent. As an example, 2-(2-methylphenyl)-2-oxazoline (8) was arylated with 4-chloroanisole (9) in toluene as solvent, yielding biphenyl (10) regioselectively as the sole product in 91% isolated yield (Scheme 5) [9].

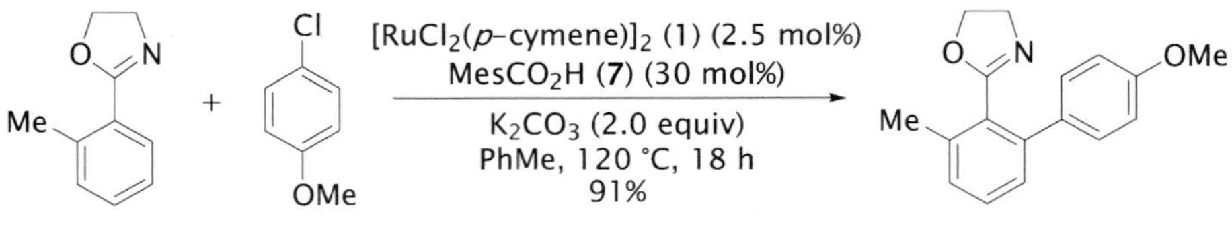

Scheme 5 Ruthenium-catalyzed direct arylation in toluene with aryl chloride 9 as electrophile and carboxylic acid 7 as preligand (Mes = 2,4,6-Me$_3$C$_6$H$_2$).

Some preparations based on ruthenium-catalyzed arylation

Preparation of 4-*n*-butyl-1-(4′-methoxy-3-methylbiphenyl-2-yl)-1H-1,2,3-triazole (6)

Apparatus	The following procedure was carried out under a nitrogen atmosphere in a flame-dried Schlenk flask equipped with a magnetic stirring bar. Magnetic stirrer, nitrogen cylinder, connecting tube, reducing valve, safety glasses, laboratory coat, protective gloves.

Chemicals   [RuCl$_2$(*p*-cymene)]$_2$ (1), (1-Ad)$_2$P(O)H (3), 4-bromoanisole (5), K$_2$CO$_3$, Na$_2$SO$_4$, toluene, Et$_2$O, water, brine, *n*-hexane.

Attention!   Safety glasses, a laboratory coat and protective gloves must be worn at all times.

Caution!   All reactions should be carried out in a well-ventilated fume hood.

Procedure   A suspension of [RuCl$_2$(*p*-cymene)]$_2$ (1) (7.7 mg, 0.012 mmol, 2.5 mol%), (1-Ad)$_2$P(O)H (3) (16 mg, 0.05 mmol, 10 mol%), K$_2$CO$_3$ (138 mg, 1.00 mmol), 4 (108 mg, 0.50 mmol) and 4-bromoanisole (5) (140 mg, 0.75 mmol) in toluene (2 mL) is stirred under N$_2$ for 18 h at 120 °C. At ambient temperature, Et$_2$O (75 mL) and H$_2$O (75 mL) are added to the reaction mixture. The separated aqueous phase is extracted with Et$_2$O (2 × 75 mL). The combined organic layers are washed with H$_2$O (50 mL) and brine (50 mL), dried over Na$_2$SO$_4$ and concentrated in vacuum. The remaining residue is then purified by column chromatography on silica gel (*n*-hexane/EtOAc 3:1) to yield 6 as a colorless oil (136 mg, 85%).

Characterization data: ^{1}H NMR (300 MHz, CDCl$_3$): δ 7.45–7.40 (m, 1H), 7.33–7.28 (m, 2H), 6.97 (d, *J* = 8.9 Hz, 2H), 6.94 (s, 1H), 6.73 (d, *J* = 8.9 Hz, 2H), 3.74 (s, 3H), 2.64 (t, *J* = 7.4 Hz, 2H), 2.12 (s, 3H), 1.52 (quint, *J* = 7.6 Hz, 2H), 1.25-1.13 (m, 2H), 0.85 (t, *J* = 7.3 Hz,

3H) ppm. ^{13}C NMR (75.6 MHz, CDCl$_3$): δ 159.0 (C$_q$), 147.8 (C$_q$), 139.2 (C$_q$), 136.1 (C$_q$), 134.6 (C$_q$), 130.1 (C$_q$), 129.8 (CH), 129.7 (CH), 129.4 (CH), 128.1 (CH), 123.1 (CH), 113.6 (CH), 55.1 (CH$_3$), 31.4 (CH$_2$), 25.0 (CH$_2$), 21.8 (CH$_2$), 17.7 (CH$_3$), 13.7 (CH$_3$) ppm. IR (NaCl, cm^{-1}): 2957, 2932, 2861, 1611, 1516, 1480, 1468, 1252, 1179, 1039, 789. MS (EI, 70 eV): m/z 321 ([M$^+$], 3%), 292 (17%), 278 (8%), 250 (30%), 237 (10%), 180 (18%), 144 (100%), 131 (8%), 91 (43%), 65 (14%). HR-MS (ESI): m/z calcd for C$_{20}$H$_{24}$N$_3$O 322.1914, found 322.1915.

Waste Disposal Organic solvent-free aqueous layers were disposed of to the sewer system. Organic solvents and organic compounds were separately collected.

Preparation of 2-(4′-methoxy-3-methyl-biphenyl-2-yl)-4,5-dihydro-oxazole (10)

Apparatus The procedure was carried out under a nitrogen atmosphere in a flame-dried Schlenk flask equipped with a magnetic stirring bar. Magnetic stirrer, nitrogen cylinder, connecting tube, reducing valve, safety glasses, laboratory coat, protective gloves.

Chemicals [RuCl$_2$(p-cymene)]$_2$ (1), mesitylene carboxylic acid (MesCO$_2$H) (7), 2-(2-methylphenyl)-2-oxazoline (8), 4-chloroanisole (9), K$_2$CO$_3$, Na$_2$SO$_4$, toluene, Et$_2$O, water, brine, n-pentane.

Attention! Safety glasses, a laboratory coat and protective gloves must be worn at all times.

Caution! All reactions should be carried out in a well-ventilated fume hood.

Procedure A suspension of [RuCl$_2$(p-cymene)]$_2$ (1) (7.7 mg, 0.012 mmol, 2.5 mol%), MesCO$_2$H (7) (25 mg, 0.15 mmol, 30 mol%), K$_2$CO$_3$ (138 mg, 1.00 mmol), 8 (81 mg, 0.50 mmol), and 4-chloroanisole (9)

(106 mg, 0.75 mmol) in toluene (2 mL) is stirred under N_2 for 18 h at 120 °C. At ambient temperature, Et_2O (75 mL) and H_2O (75 mL) are added to the reaction mixture. The separated aqueous phase is extracted with Et_2O (2 × 75 mL). The combined organic layers are washed with H_2O (50 mL) and brine (50 mL), dried over Na_2SO_4 and concentrated in vacuum. The remaining residue is then purified by column chromatography on silica gel (*n*-pentane/Et_2O 1:1) to yield **10** as a light yellow oil (121 mg, 91%).

Characterization data: ^{1}H NMR (300 MHz, CDCl$_3$): δ 7.37–7.28 (m, 3H), 7.18 (md, *J* = 8.1 Hz, 2H), 6.90 (md, *J* = 8.8 Hz, 2H), 4.15 (t, *J* = 9.3 Hz, 2H), 3.91–3.81 (m, 5H), 2.40 (s, 3H) ppm. ^{13}C NMR (75.6 MHz, CDCl$_3$): δ 164.6 (C$_q$), 158.8 (C$_q$), 141.5 (C$_q$), 137.4 (C$_q$), 133.6 (C$_q$), 129.5 (CH), 129.4 (CH), 128.5 (CH), 128.1 (C$_q$), 127.1 (CH), 113.4 (CH), 67.1 (CH$_2$), 55.2 (CH$_3$), 55.0 (CH$_2$), 19.7 (CH$_3$) ppm. IR (KBr, cm^{-1}): 3062, 2933, 1884, 1666, 1610, 1514, 1461, 1346, 1291, 1248, 1180, 937. MS (EI, 70 eV): *m/z* (%) = 267 ([M$^+$], 17%), 266 (100%), 222 (4%), 152 (2%). HR-MS (ESI): *m/z* calcd for $C_{17}H_{17}NO_2$ 267.1259, found 267.1263.

Waste Disposal Organic solvent-free aqueous layers were disposed of to the sewer system. Organic solvents and organic compounds were separately collected.

References

1 L. Ackermann, *Modern Arylation Reactions*, Wiley-VCH, Weinheim, **2009**.

2 A. de Meijere, F. Diederich, *Metal-Catalyzed Cross-Coupling Reactions*, Wiley-VCH, Weinheim, **2004**.

3 (a) J. Hassan, M. Sevignon, C. Gozzi, E. Schulz, M. Lemaire, *Chem. Rev.* **2002**, *102*, 1359; (b) D. Alberico, M.E. Scott, M. Lautens, *Chem. Rev.* **2007**, *107*, 174.

4 I. Omae, *Coord. Chem. Rev.* **2004**, *248*, 995.

5 L. Ackermann, *Top. Organomet. Chem.* **2007**, *24*, 35.

6 For representative examples, see: (a) F. Kakiuchi, Y. Matsuura, S. Kan, N. Chatani, *J. Am. Chem. Soc.* **2005**, *127*, 5936; (b) L. Ackermann, *Org. Lett.* **2005**, *7*, 3123; (c) S. Oi, E. Aizawa, Y. Ogino , Y. Inoue , *J. Org. Chem.* **2005**, *70*,

3113; (d) L. Ackermann, R. Born, P. Álvarez-Bercedo, *Angew. Chem.* **2007**, *119*, 6482; *Angew. Chem. Int. Ed.* **2007**, *46*, 6364; and references cited therein. For phosphine ligand-free direct arylations, see: (e) L. Ackermann, A. Althammer, R. Born, *Synlett* **2007**, 2833; (f) L. Ackermann, A. Althammer, R. Born, *Tetrahedron* **2008**, *64*, 6115.

7 (a) L. Ackermann, *Synthesis* **2006**, 1557; (b) L. Ackermann, *Synlett* **2007**, 507.

8 L. Ackermann, A. Althammer, R. Born, *Angew. Chem.* **2006**, *118*, 2681; *Angew. Chem. Int. Ed.* **2006**, *45*, 2619.

9 L. Ackermann, R. Vicente, A. Althammer, *Org. Lett.* **2008**, *10*, 2299.

Part II
Solvents

13

Diaryl Disulfides from Thiols in Water–Ammonia

Mónica Carril, Raul SanMartin and Esther Domínguez

The disulfide bridge, as it is commonly known, is a fundamental covalent bond in biochemistry (both bioinorganic and bioorganic), as relevant biological processes (e.g. secondary and tertiary structure formation in proteins by linkage of cysteine aminoacid moieties and derived bacterial/eukaryotic cell mechanisms against oxidation) are based on its unique properties [1]. Apart from industrial uses, such as vulcanization of rubber or rechargeable lithium batteries [2], the use of disulfides, and in particular diaryl disulfides in organic synthesis, is specially remarkable [3]. The reversibility of the S–S bond formation makes these compounds suitable for additions to alkenes and alkynes, for substitution of aryl halides or for exchange to form unsymmetrical disulfides [4]. Typically, oxidation of thiols by a metal oxidant in a basic environment is the most widely employed method for the generation of these useful compounds [5], although other protocols (use of benzotriazolated thiols and *N*-trifluoroacetyl arenesulfenamides as precursors or oxidation by means of 1,3-dibromo-5,5-dimethylhydantoin (DBDMH, *inter alia* [6]) have been reported. In the search for more sustainable synthetic procedures that avoid the use of potentially toxic and often flammable organic solvents or expensive and sometimes harmful metal salts [5, 6], a metal-free approach conducted in aqueous media would be welcome for both scientific and industrial applications. Water is an excellent solvent for organic transformations because of its intrinsic characteristics (nontoxic, cheap, safe, nonflammable) and the usually easy isolation of the final organic products from an aqueous environment. As shown below, this aim has been fulfilled by the presented metal-free synthesis of diaryl disulfides in basic aqueous medium.

Some preparations of diaryl disulfides from thiols in water–ammonia

Preparation of diphenyl disulfide

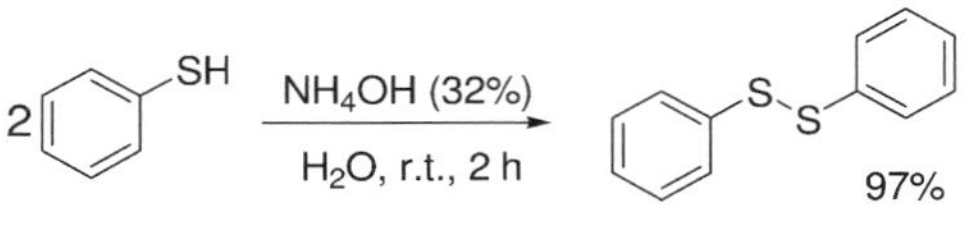

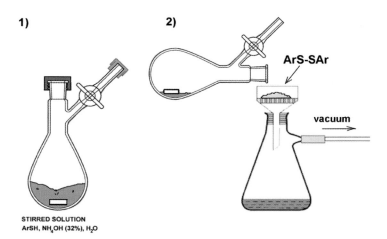

Scheme 1

Apparatus	10 mL Schlenk flask with septum, magnetic stirrer, syringes (1 mL), funnel, filter paper, safety glasses, laboratory coat, protective gloves.
Chemicals	PhSH, NH₄OH (32% aqueous solution of NH₃), H₂O.

Chemicals: PhSH, NH_4OH (32% aqueous solution of NH_3), H_2O.

Attention! Safety glasses and protective gloves must be worn at all times.

Caution! Because of their moderate toxicity and volatility, care should be taken to avoid inhalation of thiophenol and vapor from the aqueous ammonia solution or contact of these chemicals with the skin. All reactions should be carried out in a well-ventilated hood.

1)   **2)**

ArS-SAr

vacuum

STIRRED SOLUTION
ArSH, NH₄OH (32%), H₂O

Figure 1   Apparatus set up for the preparation of the diphenyl disulfide.

Procedure

The Schlenk flask is charged in an open atmosphere with thiophenol (0.1 mL, 0.94 mmol), a 32% aqueous solution of NH_3 (0.15 mL) and water (6 mL). The flask is sealed and the mixture stirred at room temperature for 2 h. The crude mixture is filtered to afford chromatographically pure diphenyl disulfide (99.7 mg, 97%) as a white powder.

Characterization data: m.p. 60–62 °C (H_2O) (Lit. 58–60 °C (EtOH) [7]). ^1H NMR (300 MHz, $CDCl_3$): δ 7.22–7.35 (6H, m, H_{arom}), 7.53 (4H, d, $J = 7.43$ Hz, H_{arom}) ppm. ^{13}C NMR (75 MHz, $CDCl_3$): δ 126.9, 127.3, 128.9, 136.8 ppm. HRMS(EI) *m/z:* Found: 218.0228. $C_{12}H_{10}S_2$ requires: 218.0224. LRMS(EI) *m/z:* 218 (M$^+$, 100%).

Preparation of di(2-naphthyl) disulfide

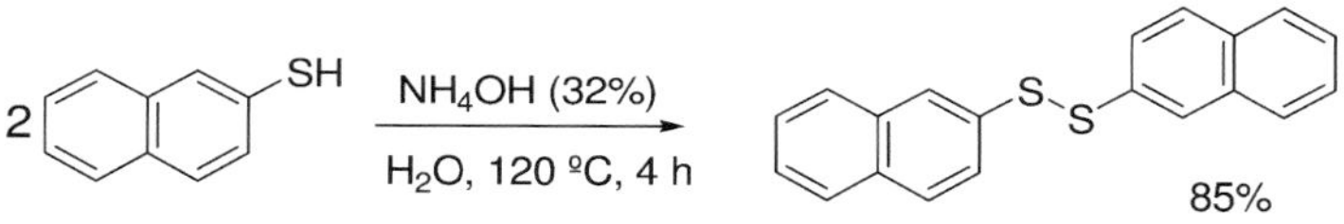

Scheme 2

Apparatus

10 mL Schlenk flask with septum, magnetic stirrer, syringes (1 mL), funnel, filter paper, safety glasses, laboratory coat, protective gloves.

Chemicals

2-Thionaphthol (naphthalene-2-thiol), NH_4OH (32% aqueous solution of NH_3), H_2O.

Attention!

Safety glasses and protective gloves must be worn at all times.

Caution!

Because of their moderate toxicity and volatility, care should be taken to avoid inhalation of 2-thionaphthol and the aqueous ammonia solution or contact of these chemicals with the skin. All reactions should be carried out in a well-ventilated hood behind a protective shield. Pressure builds up when the Schlenk flask is heated.

1) 2)

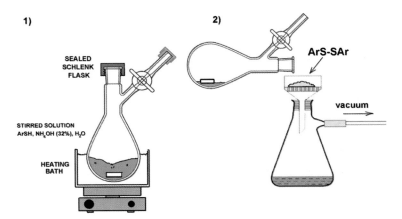

Figure 2 Apparatus set up for the preparation of the di(2-naphthyl) disulfide.

Procedure

The Schlenk flask is charged in an open atmosphere with 2-thio-naphthol (161.9 mg, 1 mmol), a 32% aqueous solution of NH$_3$ (0.15 mL) and water (6 mL). The flask is sealed and the mixture is heated to 120 °C with stirring for 4 h. After cooling, the crude mixture is filtered to afford chromatographically pure di(2-naphthyl) disulfide (134.5 mg, 85%) as a beige powder.

Characterization data: m.p. 138–140 °C (H$_2$O) (Lit. 135–138 °C (EtOAc) [8]). ^{1}H NMR (300 MHz, CDCl$_3$, Me$_4$Si): δ 7.49–7.45 (4H, m, H$_{arom}$), 7.65 (2H, d, J = 8.66 Hz, H$_{arom}$), 7.82–7.73 (6H, m, H$_{arom}$), 8.01 (2H, s, H$_{arom}$) ppm. ^{13}C NMR (75 MHz, CDCl$_3$, Me$_4$Si): δ 125.6, 126.2, 126.5, 126.7, 127.4, 127.7, 128.9, 132.4, 133.4, 134.2 ppm. HRMS(EI) *m/z:* Found: 318.0544. C$_{20}$H$_{14}$S$_2$ requires: 318.0537. LRMS(EI) *m/z:* 318 (M$^+$, 12%), 254 (16%), 159 (40%), 115 (100%).

Waste Disposal

The basic aqueous solution containing traces of unreacted thiol or disulfide should be disposed of in the aqueous waste container. The contaminated filter paper should be placed in the solid waste container.

Explanation

After assaying different water-soluble bases, it was found that aqueous ammonia is by far the most convenient one for the desired transformation (e.g. the use of NaOH and K$_2$CO$_3$ provided 31% and < 5% yields, respectively). Surprisingly, other insoluble organic bases like tetramethyl ethylenediamine or 4-dimethylaminopyridine provided slightly lower yields than ammonium hydroxide although higher temperatures are required. The generality of the method, which employs the simplest nitrogen-containing base (ammonia) is proved by the fact that the reaction outcome is not affected by the electronic nature or steric hindrance of the substituents present in the aryl or heteroaryl rings of thiols.

Regarding the reaction mechanism, we propose that the oxygen that is naturally dissolved in water is responsible for the formation of thiyl radicals, probable intermediates in this oxidation reaction with similarities to metal-catalyzed processes [5, 6]. Examples of aerobic oxidations of thiols are scarce in the literature and the reported examples require an external input of oxygen and, in some cases, even the presence of a metallic catalyst [9]. However, in our case, there is no need of an external supply of oxygen and the reaction proceeds with the oxygen already present in the water. When degassed water and an inert atmosphere (argon) were employed, no trace of target disulfide was detected, but on using water without previous degassing and an argon atmosphere, 95% of diphenyl disulfide was obtained from thiophenol.

Taking into account the extreme simplicity of the method, the very low cost and the mildness of the conditions required, and finally the advantages in terms of safety and environmental issues, the protocol presented can be defined as a green procedure [10].

References

1 (a) M. Bodanszky *Principles of Peptide Synthesis: Reactivity and Structure Concepts in Organic Chemistry,* Springer Verlag, Heidelberg, **1984**; (b) L. Ellgaard, L.W. Ruddock, *EMBO Rep.* **2005**, *6*, 28.

2 (a) A.J. Parker, N. Kharasch, *Chem. Rev.* **1959**, *59*, 583; (b) T. Maddanimath, Y.B. Khollam, M. Aslam, I.S. Mulla, K. Vijayamohanan, *J. Power Sources* **2003**, *124*, 133.

3 (a) A. Alagic, A. Koprianiuk, R. Kluger, *J. Am. Chem. Soc.* **2005**, *127*, 8036; (b) O. Higuchi, K. Tateshita, H. Nishimura, *J. Agric. Food Chem.* **2003**, *51*, 7208.

4 (a) K. Tanaka, K. Ajiki *Tetrahedron Lett.* **2004**, *45*, 5677; (b) M. Arisawa, A. Suwa, M. Yamaguchi, *J. Organomet. Chem.* **2006**, *691*, 1159; (c) T. Kondo, T.A. Mitsudo, *Chem. Rev.* **2000**, *100*, 3205; (d) N. Taniguchi, *J. Org. Chem.* **2006**, *71*, 7874; (e) B.C. Ranu, T. Mandal, *Tetrahedron Lett.* **2006**, *47*, 6911; (f) I.P. Beletskaya, C. Moberg, *Chem. Rev.* **1999**, *99*, 3435; (g) V.P. Ananikov, M.A. Kabeshov, I.P. Beletskaya, G.G. Aleksandrov, I.L. Eremenko, *J. Organomet. Chem.* **2003**, *687*, 451; (h) S. Kumar, L. Engman, *J. Org. Chem.* **2006**, *71*, 5400; (i) N. Taniguchi, *J. Org. Chem.* **2004**, *69*, 6904.

5 For procedures using iron (III) salts, see: (a) N. Iranpoor, B. Zeynizadeh, *Synthesis* **1999**, 49. Cobalt (II) salts: (b) S.M.S. Chauhan, A. Kumar, K.A. Srinivas, *Chem. Comm.* **2003**, 2348. Chromium salts: (c) S. Patel, B.K. Mishra, *Tetrahedron Lett.* **2004**, *45*, 1371. Rhodium salts: (d) M. Arisawa, C. Sugata, M. Yamaguchi, *Tetrahedron Lett.* **2005**, *46*, 6097.

Molybdenum salts: (e) R. Sanz, R. Aguado, M.R. Pedrosa, F. Arnáiz, *Synthesis* **2002**, 856.

6 (a) M.M. Hashemi, H. Ghafuri, Z. Karimi-Jaberi, *J. Sulfur Chem.* **2006**, *27*, 165; (b) B. Karami, M. Montazerozohori, M.H. Habibi, *Molecules* **2005**, *10*, 1358; (c) R. Hunter, M. Caira, N. Stellenboom, *J. Org. Chem.* **2006**, *71*, 8268; (d) S. Antoniow, D. Witt, *Synthesis*, **2007**, 363; (e) M. Bao, M. Shimizu, *Tetrahedron* **2003**, *59*, 9655; (f) A. Khazaei, M.A. Zolfigol, A. Rostami, *Synthesis* **2004**, 2959.

7 G. De Martino, G. La Regina, A. Coluccia, M.C. Edler, M.C. Barbera, A. Brancale, E. Wilcox, E. Hamel, M. Artico, R. Silvestri, *J. Med. Chem.* **2004**, *47*, 6120.

8 S.C. Banfield, A.T. Omori, H. Leisch, T. Hudlicky *J. Org. Chem.* **2007**, *72*, 4989.

9 (a) M. Kirihara, K. Okubo, T. Uchiyama, Y. Kato, Y. Ochiai, S. Matsushita, A. Hatano, K. Kanamori, *Chem. Pharm. Bull.* **2004**, *52*, 625; (b) T.J. Wallace, A. Schriesheim, W. Bartok, *J. Org. Chem.* **1963**, *28*, 1311.

10 M. Carril, R. SanMartin, E. Domínguez, I. Tellitu, *Green Chem.* **2007**, *9*, 315.

14

Electrochemical Synthesis of Polypyrrole Employing Green Chemistry Principles

Anthony Fusco and Patricia Ann Mabrouk

Polypyrrole is an example of a conducting polymer, so called from its ability to conduct electricity. Conducting polymers have many uses including LED displays, as an anti-static coating in photocopying, and corrosion protection [1, 2]. Conducting polymers such as polypyrrole can be synthesized chemically using a chemical oxidant or electrochemically. Traditionally, when polymers such as polypyrrole are synthesized electrochemically, concentrated acid or organic solvents are used as solvents. This is undesirable as these solvents require expensive disposal methods and represent a potential threat to the environment. In this experiment, students synthesize polypyrrole using only the monomer and an electrolyte. This approach is both cleaner and safer. In the laboratory, students will synthesize a sample of polypyrrole via cyclic voltammetry and then test the electrical properties of the polypyrrole using two different methods. First, a voltmeter will be used to demonstrate that the polymer conducts electricity. The second method is more hands on and will allow the students to be creative. In this part, students will create a simple electrical circuit using batteries, an LED light and their sample of polypyrrole. By completing this project, students will gain an appreciation of the principles of green chemistry and will have hands on experience in electrochemical polymer synthesis and an appreciation of the unique electrical properties of polypyrrole [1–3].

Preparation of polypyrrole using green electrochemistry

Apparatus   10 mL volumetric flask, 10 mL beaker, indium tin oxide on glass slide (Delta Technologies, R_s 4-8 Ω; 7 mm $\times$ 50 mm $\times$ 0.5 mm; CG-41INCUV) , Ag/AgCl reference electrode or other suitable reference electrode, Pt mesh (or stainless steel) counter electrode, 3M™ (double sided foam) heavy duty mounting tape, digital voltmeter,

LED, battery holder and 3 AA batteries, tweezers, Kimwipes®, electronic balance, potentiostat, Faraday cage (if available), safety glasses, laboratory coat, disposable gloves.

Chemicals   Pyrrole (Aldrich, 98%) should be freshly distilled. Depending on the time available, the students can set up and distill pyrrole the week before the actual experiment. Tetra-*n*-butylammonium perchlorate (Alfa-Aesar) should be purchased and can be used »as is« as the supporting electrolyte.

Attention!   Laboratory coat, gloves and safety glasses should be worn at all times.

Caution!   Pyrrole should be handled in a chemical fume hood. Care should be taken to avoid inhaling pyrrole or contact with skin.

Procedure   Students can work individually or as teams (2–3 students) on this project. In the laboratory, students are given a 2 mL volumetric flask, tetra-*n*-butylammonium perchlorate, distilled pyrrole, and a 10 mL beaker. These are used to prepare a 0.3 M tetra-*n*-butylammonium perchlorate (TBAP) solution in pyrrole in the 2-mL volumetric flask (in the chemical fume hood). The students are also provided with an indium tin oxide on glass (ITO) slide (coated with ITO on one side), which will serve as the working electrode, an Ag/AgCl reference electrode, and a Pt mesh counter electrode (Pt mesh provides a large surface area). If the Ag/AgCl reference and/or Pt mesh counter electrodes are not available, a silver wire can be used as a quasi-reference electrode and a piece of stainless steel can be used as a counter electrode. The piece of steel should be at least as large as the ITO slide if not greater.

Next, the students will set up the electrochemical cell. They add all the TBAP/pyrrole solution to the 10-mL beaker, which serves as the electrochemical cell. Half of the ITO slide should be covered with the pyrrole once the slide is placed vertically in the solution. The ITO slide should be carefully handled with tweezers because oils from the students' fingers can inhibit the deposition of the polymer on the ITO slide. Next, students add the counter electrode and reference electrode. The cell containing the TBAP/pyrrole solution and the working, counter, and reference electrodes can now be placed in the Faraday cage, if available. The Faraday cage electrically shields the cell from stray electrical noise. Students should make the appropriate contacts to the correct electrodes (working, reference, and counter). Students should also ensure that none of the electrodes are touching each other (electrical short) or that the clip used to contact the ITO/glass electrode does not touch the TBAP/pyrrole solution (see Figure 1).

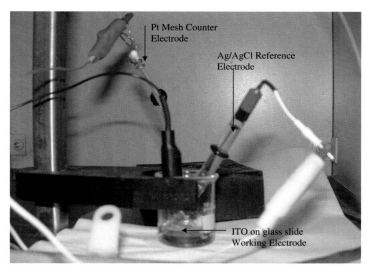

Figure 1   Set-up of electrochemical cell with counter, reference, and working electrodes in the Faraday cage.

Once the cell is set up and all electrodes are properly contacted, students should program the potentiostat as follows. The potential for the working electrode should be varied between −1000 and +1500 millivolts versus Ag/AgCl at a scan rate of 100 mV s^{-1} for a minimum of 30 cycles. The run should take approximately 25 min to complete. During the run, the students should observe the cell and make note of any changes that take place concerning the ITO/glass working electrode. Over time, as the oxidative polymerization of pyrrole takes place, a brown film will form on the working electrode.

When the experiment is finished, students should carefully unclip the electrodes and remove the cell from the Faraday cage. The students should remove the working electrode using tweezers and observe the formation of a solid black polymer on one side of the ITO glass. Once the ITO slide has been removed, holding the ITO/glass slide so that the polymer is at the bottom of the slide, the students

Figure 2 Photograph showing the proper procedure for rinsing the polypyrrole with squirt bottle and tweezers. The bottle is angled so that water drips down the glass and completely rinses the polymer. Tweezers are used to keep the working electrode clean.

should rinse the slide carefully. This should be done by aiming a stream of distilled or de-ionized water from a squirt bottle at the top of the slide which has no polymer on it and allowing the water to drain onto the polymer at the bottom of the slide. Students should angle the bottle to ensure that the polymer does in fact get rinsed. They should also be sure not to directly squirt water at the polymer film as otherwise this film will fall off the slide (see Figure 2). To dry the polymer, students should fold a Kimwipe®, contact one corner of the polymer with the Kimwipe®, and allow capillary action to absorb the water. This should be repeated until most of the water is gone (see Figure 3). The polymer-coated ITO/glass slide should be allowed to rest (polymer side up) on a paper towel on a laboratory bench for 1 min.

Next, students should roll a piece of heavy-duty mounting tape (about 2.5 cm long) onto the slide so that every part of the polymer makes contact with the tape (see Figure 4). The students should apply pressure all over the slide so that the entire polymer sticks to the tape. To separate the slide from the tape, students should repeat the same rolling motion.

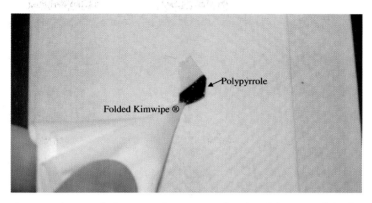

Figure 3 Photograph illustrating the suggested method of drying the polypyrrole with a folded Kimwipe®. Using the edge of the wipe will dry the polymer while keeping it intact on the ITO glass slide.

Once students have a thick, continuous polymer on a piece of tape, they should proceed to test the electrical properties of the polypyrrole. First, the students should use the probes of an ordinary voltmeter to contact the sample and observe the resistance across the polymer. Once students understand that the polymer is conductive, they should proceed to create a circuit using batteries, a battery holder, an LED, and the polymer.

While students can create the circuit in whichever manner they like, the best approach to creating the circuit is as follows. Students

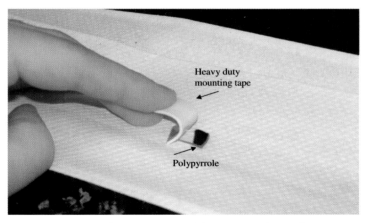

Figure 4 Photograph demonstrating the procedure for transferring the polypyrrole from the ITO-coated glass slide to a piece of heavy-duty mounting tape by rolling the tape onto the slide.

should connect one end of one of the wires/contacts from the battery holder to a wire attached to the LED. Then using the other wire from the LED and the other contact from the battery holder, the students should make contact with the polymer in two different locations. This ensures that there is solid contact with the polymer, thus minimizing potential problems. Contact and completion of the circuit should result in the illumination of the LED, thus proving the electrical conductivity of polypyrrole.

Waste Disposal The TBAP/pyrrole solution can be recycled and used many times by successive groups of students. If the instructor chooses, students should dispose of the TBAP/pyrrole solution in a waste disposal container.

Explanation

Cyclic voltammetry is a fast and effective way to create a film of polypyrrole [3]. On the first scan, the pyrrole becomes oxidized. As the concentration of oxidized pyrrole increases, the oxidized pyrrole forms dimers, oligomers, and finally a polymer which precipitates onto the ITO-coated side of the working electrode.

By completing the experimental procedure, students should understand that polypyrrole is a good electrical conductor. The students should contact the polymer film in two places using the probes of a standard digital voltmeter operating in the resistance mode. The resistance of the polypyrrole films obtained experimentally averaged 118.2 KΩ with a standard deviation of 4.35 KΩ (n = 3). At this point, the instructor may wish to lead a discussion introducing the traditional methods of conducting polymer synthesis and explaining that conducting polymers are usually synthesized in acids or toxic organic solvents. This will enable students to appreciate why this particular method of synthesizing polypyrrole is more environmentally benign.

References

1 T.A. Skotheim, J.R. Reynolds, *Handbook of Conducting Polymers: Conjugated Polymers. Processing and Applications. 3rd ed.*; CRC Press: Boca Raton, **2007**.

2 M. Angelopoulos, IBM *J. Res. Dev.* **2001**, *45*, 57.

3 P.H. Rieger, *Electrochemistry. 2nd ed.*; Chapman & Hall, Inc.: New York, **1994**.

15

Ammonia-Sensing Cyanoaurate Coordination Polymers

Daniel B. Leznoff, Michael J. Katz and Julie Lefebvre

Polymers result from the assembly of molecular building blocks. In a coordination polymer, these building blocks are composed of inorganic metal ions and bridging ligands, which are held together by metal-ligand coordinative bonds. The metal atoms in coordination polymers can be imagined as assembling points for the bridging ligands. The combination of the geometry of the ligands and the coordination sphere of the metal controls the type of structure formed. Coordination polymers are synthesized using a modular self-assembly approach, in which the metal cation, bridging ligand and, in some cases, capping ligand building blocks can all be independently chosen to achieve a particular structure and property [1]. Given their ease of synthesis and unlimited design potential, coordination polymer research has rapidly grown in recent years, targeting the design of new functional materials with tunable conductive, magnetic, optical, or porous properties [2].

The oldest coordination polymer has a rich history. This is Prussian Blue, $Fe_4^{III}[Fe^{II}(CN)_6]_3 \cdot 14H_2O$, which was first reported in 1704 and was used as a blue dye in Europe starting in the 1700s [3]. It was exported to Japan in the 1850s, transforming woodblock printing with its vivid color, and is still used as the backing for »carbon paper«. More recently, the Prussian Blue structure type, which has a three-dimensional cubic array of M-CN-M' bridging units [4], has been explored for its magnetic, porous, and nanoscopic properties. Other *cyanometallate* building blocks, $[M(CN)_x]^{n-}$ (x = 4,6) have also been widely utilized in coordination polymer synthesis [5].

The linear $[M(CN)_2]^-$ units (M = Ag, Au) have some unique features relative to other *cyanometallate* anions. In particular, d^{10} gold(I) ions are often attracted to each other; this effect has been termed *aurophilicity* [6]. These attractive aurophilic interactions have a strength of an order of magnitude similar to that of hydrogen bonding. $[Au(CN)_2]^-$ anions often aggregate via these aurophilic interactions. In addition, materials that contain aurophilic interactions are often strongly luminescent [7]. The ability of [Au(CN)$_2$]$^-$ to influence the supramolecular structure via aurophilic interactions and simultaneously generate materials that are potentially luminescent has recently been exploited in coordination polymer design and synthesis [8].

The following experiment demonstrates the preparation of [Au(CN)$_2$]$^-$-containing coordination polymers that show vapochromic and vapoluminescent properties. Vapochromic materials have potential applications as chemical sensors as they display optical absorption changes when exposed to specific gases or vapors of volatile organic compounds (VOCs), while vapoluminescent materials change emission properties upon exposure to a particular analyte. Of course, these sensor materials are environmentally important since they allow us to sense, detect and record the current status of gases and vapors in the atmosphere.

These experiments will expose students to the »green chemical« concepts of small-scale reactions and solvent-free synthesis, as well as enabling them to prepare materials that can be used to sense toxic ammonia in the environment.

Preparation and properties of cyanoaurate coordination polymers

Preparation of metal-cyanoaurate(I) coordination polymers

Apparatus	Two 25 mL glass beakers, magnetic stirrer and two micro-stirbars, filter paper and Hirsch funnel, graduated cylinder (10 mL), analytical balance, Pasteur pipettes, small spatula, small agate mortar and agate pestle, long UV-wavelength lamp or box, safety glasses, laboratory coat, protective gloves. (Optional: IR spectrometer, solid state visible reflectance spectrometer, fluorimeter).
Chemicals	Cu(NO$_3$)$_2 \cdot$6H$_2$O, Zn(NO$_3$)$_2 \cdot$6H$_2$O, K[Au(CN)$_2$], 50 mL H$_2$O.
Attention!	Safety glasses and protective gloves must be worn at all times.
Caution!	Metal cyanides should not come into contact with strong acids. Although [Au(CN)$_2$]$^-$ is less toxic than other cyanide-containing

species, care should still be taken in handling it. Gloves should be worn at all times.

Procedure

(a) **Aqueous synthesis of Cu[Au(CN)$_2$]$_2$(H$_2$O)$_2$**

Prepare a pale blue aqueous solution (5 mL) of Cu(NO$_3$)$_2$·6H$_2$O (15 mg, 0.05 mmol) in a 25 mL beaker. In a second 25 mL beaker, prepare a 5 mL colorless aqueous solution of KAu(CN)$_2$ (30 mg, 0.10 mmol). With stirring, add the cyanoaurate solution to the copper(II) solution using a Pasteur pipette. A pale green powder of Cu[Au(CN)$_2$]$_2$(H$_2$O)$_2$ forms immediately. After stirring for 5 min, filter the precipitate onto a Hirsch funnel and filter paper. Do not pour the slurry into the funnel directly, but instead take up the suspension into a Pasteur pipette and add it slowly drop by drop onto the center of the filter paper—this will increase the yield. Finally, air dry the solid and obtain the yield. IR and solid state visible reflectance spectra of the product can be measured.

Characterization data: The yield of Cu[Au(CN)$_2$]$_2$(H$_2$O)$_2$ is nearly quantitative. Thermogravimetric analysis: –2H$_2$O at 180 °C; dec. 320 °C. IR (KBr, cm^{-1}): 3246 (m), 2217 (s), 2194 (vw), 2172 (s), 1633 (w). Solid-state visible spectrum: 535 nm.

(b) **Solvent-free synthesis of Zn[Au(CN)$_2$]$_2$**

Weigh out KAu(CN)$_2$ powder (30 mg, 0.10 mmol) and transfer it into a small agate mortar. Separately, weigh Zn(NO$_3$)$_2$·6H$_2$O (15 mg, 0.05 mmol) and add it to the same mortar. Put the mortar under a long UV-wavelength lamp and, with the lamp on, grind the two colorless solids together in the mortar with the pestle for 3 min to form a wet paste. Note the large visible emission change that occurs upon solid-state reaction! Put the mortar into the oven for 10 min to dry the paste. This mixture also contains KNO$_3$, which can be washed out with water if a pure sample is desired, but this is not necessary for the reaction with ammonia described below. IR and emission spectra of the product can be measured.

Characterization data: IR (KBr, cm^{-1}): 2216 (w), 2198 (s), 2158 (w), 517(m). m.p. 270 °C (decomp.). Solid state emission spectrum (nm): 390, 480 (excitation at 345).

Reaction of metal cyanoaurate coordination polymers with gaseous ammonia

Apparatus   Small vial and cap, small strip of filter paper, Pasteur pipette and bulb, safety glasses, laboratory coat, protective gloves, UV lamp or box. (Optional: IR spectrometer, solid state visible reflectance spectrometer, fluorimeter).

Attention!   Safety glasses and protective gloves must be worn at all times.

Caution!   Conduct this reaction in a well-ventilated fume hood since ammonia gas is toxic and noxious.

Chemicals $Cu[Au(CN)_2]_2(H_2O)_2$ and $Zn[Au(CN)_2]_2$ as prepared above, 10 mL of concentrated ammonia (approx. 29%) in a sealed 50-mL bottle.

Procedure   Place green $Cu[Au(CN)_2]_2(H_2O)_2$ powder (5–10 mg) into a small vial. Fix the bulb on the pipette and uncap the ammonia bottle *in a fume hood*. Using the pipette, take one drop of ammonia and put it on the thin strip of filter paper. Quickly hang the paper strip inside the vial containing the powder and hold it in place by pressing the vial cap down onto the top of the paper (see Figure 1). The color of the powder rapidly changes to a dark blue, indicative of ammonia binding. This color change and the associated IR band changes act as a sensory response for ammonia gas. An IR and visible reflectance spectrum of this product can be obtained.

To test the vapoluminescent response of the colorless $Zn[Au(CN)_2]_2$ still in the mortar (from above), first put the mortar

Figure 1 Experimental set-up of sealed vial with solid $Cu[Au(CN)_2]_2(H_2O)_2$ and ammonia soaked filter paper strip.

with the white powder under a long UV-wavelength lamp. Fix the bulb on a new pipette and uncap the ammonia bottle *in a fume hood.* Using the pipette, pull approximately 2 mL of ammonia *gas* from the head space of the bottle. Do not suck up any liquid into the pipette. Place the tip of the pipette close to the white $Zn[Au(CN)_2]_2$ in the mortar and spray the ammonia gas onto it. Upon adding the ammonia vapor, the luminescent emission color changes dramatically (if you watch carefully, you can see two changes in emission!). After 2 min, take the mortar out of the lamp and observe the bright yellow color of the product. The emission spectrum of the final $Zn[Au(CN)_2]_2(NH_3)_2$ product can be measured using a fluorimeter (excitation energy 400 nm) and compared with the emission of the starting material. What happens if you place the Zn– and Cu–ammonia compounds in the oven for 10 min?

Characterization data for $Cu[Au(CN)_2]_2(NH_3)_x$: Visible reflectance: 435 nm. IR (KBr, cm^{-1}): 2141, 2136.

Characterization data for $Zn[Au(CN)_2]_2(NH_3)_2$: IR (KBr, cm^{-1}): 3290 (m), 3178 (w), 2158 (s), 2117 (sh), 1202 (s), 618 (br,m). Emission (nm): 50 (excitation at 400).

Waste Disposal The solid coordination polymers should be disposed of as solid waste in an appropriately designated »chemical solid waste« container. The small volumes of aqueous wastes do not normally require special treatment but may be disposed of as »liquid chemical waste« if desired.

Explanation

The synthesis of $Cu[Au(CN)_2]_2(H_2O)_2$ is typical of most coordination polymers: it is very simple and uses environmentally benign water as a solvent. The synthesis of the zinc(II) analog employs »mechanochemistry«, which does not require any solvent at all and is thus very environmentally friendly [9]. Although nonporous, the structures of both compounds [10] are surprisingly flexible in the solid state [11]; they can dynamically adapt to accommodate external analytes such as ammonia with a concomitant change in their physical properties that act as a sensory readout.

Thus, the $Cu[Au(CN)_2]_2(H_2O)_2$ system is a vapochromic material, displaying an optical absorption change upon exposure to ammonia (see Figure 2) and se-

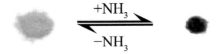

Figure 2 Reversible reaction of vapochromic $Cu[Au(CN)_2]_2(H_2O)_2$ with NH_3 vapor and the associated color changes.

lected other analytes. The color change can be attributed to the ammonia molecules binding to the copper(II) center, expelling the bound H_2O, and thereby increasing the crystal field splitting. As a consequence, the color of the vapochromic $Cu[Au(CN)_2]_2(H_2O)_2$ changes as the d-d absorption bands shift with the donor. In addition, the N-cyano groups bound to the copper(II) center are also affected by the change in the copper coordination sphere, and the changes in the CN absorption ($2100–2220$ cm^{-1}) can be used to sense the presence of ammonia as well [12].

The nature of the vapochromism of the $Zn[Au(CN)_2]_2$ system, in which the emission energy changes upon exposure to ammonia (see Figure 3), is more complicated. As in the copper(II) analog, the ammonia molecules bind to the zinc(II) center, but the d^{10} zinc(II) is colorless, hence there is no d-d absorption band to shift. However, the overall superstructure adapts to accommodate the ammonia molecules, changing the strength of the aurophilic interactions in the system. Since the emission source is the aurophilic bonding present in the structure of $Zn[Au(CN)_2]_2$, this structural change causes a change in emission energy [13].

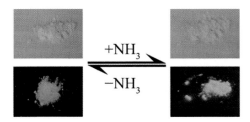

Figure 3 Reversible reaction of vapoluminescent $Zn[Au(CN)_2]_2$ with NH_3 vapor and the associated visible color (top) and emission (bottom) changes.

Ammonia detectors have a variety of applications in agriculture, the automotive industry, industrial refrigeration, medical diagnosis, and anti-terrorism [14]. From a health perspective, the human nose is capable of sensing ammonia at a concentration of 50 ppm, but the permissible long term exposure limit of ammonia is below this, at 20 ppm. For these reasons, the design of materials for the detection of ammonia is of great interest. The easily observable color change for the $Cu[Au(CN)_2]_2(H_2O)_2$ upon NH_3 binding is a useful sensory response. The $Zn[Au(CN)_2]_2$ coordination polymer's emission sensory pathway, while less direct than in the case of the copper(II) analog, is actually more sensitive, with detection limits in the low ppb for NH_3. Thus, both of these coordination polymers can act as ammonia sensors [15].

References

1 (a) P.J. Steel, *Acc. Chem. Res.* **2005**, *38*, 243; (b) D. Braga, F. Grepioni, A.G. Orpen, *Crystal Engineering: From Molecules and Crystals to Materials*, Kluwer Academic Publishers: Dordrecht, **1999**.

2 (a) C. Janiak, *Dalton Trans.* **2003**, *14*, 2781; (b) M. Eddaoudi, D.B. Moler, H. Li, T.M. Reineke, M. O'Keeffe, O.M. Yaghi, *Acc. Chem. Res.* **2001**, *34*, 319; (c) L. Ouahab, *Chem. Mater.* **1997**, *9*, 1909.

3 Anonymous, *Miscellanea Berolinensia ad Incrementum Scientiarum (Berlin)* **1710**, *1*, 377.

4 (a) J.F. Keggin, F.D. Miles, *Nature* **1936**, *137*, 577; (b) H.J. Buser, A. Ludi, W. Petter, D. Schwarzenbach, *J. Chem. Soc., Chem. Commun.* **1972**, 1299.

5 (a) J. Lefebvre, D.B. Leznoff, In *Macromolecules Containing Metal and Metal-Like Elements* (A.S. Abd-El-Aziz, C.E. Carraher Jr., C.U. Pittman Jr., M. Zeldin, eds.), Wiley-Interscience **2005**, *5*, 155; (b) K.R. Dunbar, R. A. Heintz, *Prog. Inorg. Chem.* **1997**, *45*, 283.

6 (a) H. Schmidbaur, *Gold Bulletin* **2000**, *33*, 3; (b) P. Pyykkö, *Angew. Chem.* **2004**, *116*, 4512; *Angew. Chem. Int. Ed.* **2004**, *43*, 4412; (c) M. Bardaji, A. Laguna, *J. Chem. Educ.* **1999**, *76*, 201.

7 A.L. Balch, *Struct. Bond.* **2007**, *123*, 1.

8 (a) D.B. Leznoff, J. Lefebvre, *Gold Bulletin* **2005**, *38*, 47; (b) M.J. Katz, P.M. Aguiar, R.J.

Batchelor, A. Bokov, Z.-G. Ye, S. Kroeker, D.B. Leznoff, *J. Am. Chem. Soc.* **2006**, *128*, 3669; (c) M.J. Katz, H. Kaluarachchi, R.J. Batchelor, A.A. Bokov, Z.-G. Ye, D.B. Leznoff, *Angew. Chem.* **2007**, *119*, 8960; *Angew. Chem. Int. Ed.* **2007**, *46*, 8804; (d) D.B. Leznoff, B.-Y. Xue, B.O. Patrick, V. Sanchez, R.C. Thompson, *J. Chem. Soc. Chem. Comm.* **2001**, 259; (e) D.B. Leznoff, B.-Y. Xue, R.J. Batchelor, F.W.B. Einstein, B.O. Patrick, *Inorg. Chem*, **2001**, *40*, 6026.

9 A.L. Garay, A. Pichon, S.L. James, *Chem. Soc. Rev.* **2007**, *36*, 846.

10 (a) J. Lefebvre, F. Callaghan, M.J. Katz, J.E. Sonier, D.B. Leznoff, *Chem. Eur. J.* **2006**, *12*, 6748; (b) B.F. Hoskins, R. Robson, N.V.Y. Scarlett, *Angew. Chem.* **1995**, *107*, 1317; *Angew. Chem. Int. Ed. Engl.* **1995**, *11*, 1203.

11 K. Uemura, R. Matsuda, S. Kitagawa, *J.Solid State Chem.* **2005**, *178*, 2420.

12 J. Lefebvre, R.J. Batchelor, D.B. Leznoff, *J. Am. Chem. Soc.* **2004**, *126*, 16117.

13 M.J. Katz, T. Ramnial, H.–Z. Yu, D.B. Leznoff, *J. Am. Chem. Soc.* **2008**, *130*, 10662.

14 B. Timmer, W. Olthuis, A. van den Berg, *Sensors and Actuators B* **2005**, *107*, 666.

15 J. Lefebvre, M.J. Katz, D.B. Leznoff, US patent #11577299; European patent # 05709022.0.

16

Aqueous Biphasic Systems for Liquid–Liquid Separations

Tameka L. Shamery, Jonathan G. Huddleston, Ji Chen, Scott K. Spear and Robin D. Rogers

Solvent extraction (SX) has numerous uses, such as cleaning samples by separating desired components from contaminants, transferring solutes from their original solvent to one that gives better results with certain instruments, and concentrating dilute solutes [8]. Traditionally, liquid–liquid solvent extraction is a widespread separation technique with applications in the analytical and organic chemistry laboratory [11]. Extraction typically involves allowing a mixture's components to partition between aqueous and nonpolar organic solutions. Liquid–liquid extraction involves the distribution of a solute between two immiscible liquid phases in contact with each other [2].

Liquid–liquid extraction uses toxic solvents such as benzene and toluene [10]; however, efforts are being made to replace toxic with nontoxic solvents. Aqueous Biphasic Systems (ABS) are a nonflammable and environmentally friendly choice because they are composed of approximately 80% water and contain no organic solvents [3, 4]. Not only are ABS systems environmentally benign and toxicologically safe, but they are available in bulk and inexpensive.

ABS systems are formed by the addition of two (or more) water soluble polymers, or a polymer and salt, to aqueous solution above the critical concentrations [3]. In this way two (or more) wholly aqueous phases are formed without the use of volatile organic compounds. The basis of phase formation in polymer–polymer ABS is incompletely understood at present but has been described thermodynamically on the basis of unfavorable segment interactions of polymers overcoming the entropy increase involved in phase separation. However, polymer–salt systems seem to phase separate on the basis of the competition for hydration between the polymer and the salt, resulting in increasing dehydration of the polymer, polyethylene glycol (PEG) chain, and phase separation. The concentration of polymer and salt required to bring about a phase separation is dependent on the polymer molecular mass (and type of polymer if polymers other than PEG are used) and the salting out strength of the salt. The phase separation of these polymers is most strongly promoted by anions having large negative Gibbs free energies of hydration [3, 7].

Exploration of the partitioning of dyes in ABS has intrinsic value in understanding the basic chemical principles responsible for interactions of molecules in such systems [5, 6, 9]. The broad category of molecules known as dyes contain many different structures. If the behavior of these various types of dyes can be understood, based on their structures, they can serve as models. From models it would be possible to predict similar compounds without excessive experimentation. Predictive models such as these would be of great value to the separations community as a whole. Propyl Astra Blue Iodide (PABI) is a porphyrin-like dye with many characteristics of interest; it is used in the textile industry and biological technologies [1].

Partitioning of PABI in aqueous biphasic systems

Apparatus   Vortex mixer, analytical balance (Denver Instruments, M-220D), centrifuge (optional), plastic transfer pipettes, micropipette and pipette tips (100–1000 µL), small glass bottle with lids (~10–20 mL), round bottle test tubes with lids, safety glasses, laboratory coat, protective gloves.

Chemicals   40% (w/w) poly(ethylene glycol) MW 2000 (reagent grade) in deionized water, 40% (w/w) potassium phosphate (K_3PO_4) in deionized water, 15% (w/w) potassium dihydrogen phosphate (KH_2PO_4) in deionized water, 10% (w/w) Propyl Astra Blue Iodide in deionized water.

Attention!   Safety glasses and protective gloves must be worn at all times when handling chemicals.

Caution!   Because of the caustic nature of the phosphate salt solutions, care should be taken to avoid direct skin or eye contact with these reagents. If exposure does occur, wash with copious amounts of water and seek medical attention if irritation develops or persists.

Procedure

1. For each partitioning condition to be tested, add 4 mL of 40% (w/w) PEG-2000 followed by 100 μL of 10% (w/w) dye solution to a test tube. Vortex mix the PEG-dye solution thoroughly.
2. Add 4 mL of the 40% (w/w) potassium phosphate stock solution to the test tube containing the PEG-dye mixture. Mix the PEG-salt solution thoroughly and allow the mixture to settle for several minutes or centrifuge, if available, for about 2 min.
3. Note the phase to which the dye partitions.
4. Using the transfer pipette, extract the upper PEG phase from the test tube and place in a clean test tube.
5. Add 4 mL of the 15% (w/w) potassium dihydrogen phosphate stock solution to the PEG solution. Mix thoroughly and allow the mixture to settle for several min or centrifuge, if available, and let stand for about 2 min.
6. Note the phase to which the dye partitions.

Waste Disposal   Whatever cannot be saved for recovery or recycling should be disposed of in a properly labeled container and sent to an appropriate and approved waste disposal facility. If at all possible, PEG-rich and salt-rich phases should be separated and disposed of in separate containers. Dispose of container(s) and unused contents in accordance with appropriate government or local requirements and regulations.

Explanation

Propyl Astra Blue Iodide (PABI), shown in Scheme 1, is a porphyrin-like dye with many characteristics of interest [1]. One of these is the ability to shift partitioning under certain conditions. It has been previously theorized that PABI is cationic, and pH might be responsible for its unique behavior [6]. However, it has also been argued that pH strength is not a predominant factor in the partitioning of PABI. PABI has shown the ability and arguably the preference to partition to the lower salt rich phase of many ABS systems [9].

Two systems that would shift in pH partitioning were chosen: a 43% (w/w) K_3PO_4/40% (w/w) PEG-2000 system and a 15% (w/w) KH_2PO_4/40% (w/w) PEG-2000 system. When the PEG from the K_3PO_4 system is extracted and added to KH_2PO_4 the dye partitioning shifts from the PEG phase to the salt phase. Because under normal conditions PEG and KH_2PO_4 will not form a biphasic system it was theorized that there must be a change in species, per-

Scheme 1 *n*-Propyl Astra Blue Iodide (PABI).

haps to K_2HPO_4. By combining solutions of $[PO_4]^{3-}$ and $[H_2PO_4]^-$, an acid dissociation equilibrium will be established to include $[HPO_4]^{2-}$. This system caused partitioning in the K_3PO_4-PEG system to shift to the salt phase.

Questions

1. What is the major difference between these two salt systems?
2. Relate the advantages/disadvantages of ABS to organic SX taking into account Green Chemistry Principles.
3. How would recovery of PEG or salt be accomplished?

Acknowledgments

TLS would like to thank The University of Alabama Department of Chemistry and the National Science Foundation for providing this Research Experience for Teachers (RET) opportunity.

References

1 F.G. Green, *The Sigma-Aldrich Handbook of Stains, Dyes and Indicators* Aldrich Chemical Company, Inc.: Milwaukee, **1990**.

2 P. Guha, R. Guha, *J. Chem. Educ.* **1992**, *69*, 73.

3 J.G. Huddleston, H.D. Willauer, S.T. Griffin, R.D. Rogers, *Ing. Eng. Chem. Res.* **1999**, *38*, 2523.

4 J.G. Huddleston, S.T. Griffin J. Zhang, H.D. Willauer, R.D. Rogers, In *Aqueous Two-Phase Systems: Methods in Biotechnology,* Vol. 11, R. Hatti-Kaul (Ed.), Humana: New Jersey, **2000**, pp. 77-94.

5 J.G. Huddleston, H.D. Willauer, K.R. Boaz, R.D. Rogers, *J. Chromatogr. B* **1998**, *711*, 237.

6 C. Ingenito, J.G. Huddleston, R.D. Rogers, An exploration of various textile dyes in aqueous biphasic systems. University of Alabama, **1997**.

7 R.D. Rogers, M.A. Eiteman, (Eds.), *Aqueous Biphasic Separations: Biomolecules to Metal Ions.* Plenum, New York, **1995**.

8 J. Rydberg, C. Musikas, G.R. Choppin, Principles and Practices of Solvent Extraction. **1992**, Marcel Dekker, Inc: New York.

9 T.L. Shamery, J.G. Huddleston, R.D. Rogers, A teaching module for liquid–liquid solvent extraction in aqueous biphasic systems. The University of Alabama, **2002**.

10 J. Sherman, B. Chin, P.D. Huibers, R. Garcia-Valls, T.A. Hatton, *Environ. Health Perspect.* **1998**, *106*, 253.

11 D.E. Turner *J. Chem. Educ.* **1994**, *71*, 173.

17

Clean, Fast Palladium-Catalyzed Reactions Using Microwave Heating and Water as a Solvent: the Heck Coupling as an Example

Cynthia McGrowan and Nicholas E. Leadbeater

Traditionally, organic chemists heat their reaction mixtures on a hotplate or in an oil or hot-water bath. These are relatively slow and inefficient ways of transferring heat to a sample because they depend on convection currents and the thermal conductivity of the reaction mixture. Also, the walls of the reaction vessel can be hotter than the contents. These thermal gradients mean that reagents or products can be decomposed over time. In addition, reactions can often take a long time to perform. By moving to microwave heating, reaction times can be dramatically shortened and product yields can be higher. Shortening the duration of known reactions is not the only advantage that microwave heating is having. It is impacting on modern organic chemistry by opening up routes to compounds that were previously not accessible. It is also a cleaner way to do preparative chemistry.

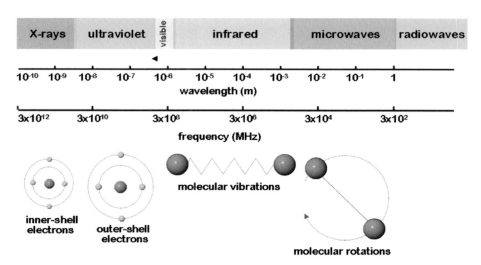

Figure 1 The electromagnetic spectrum.

Microwave irradiation is of relatively low energy. The microwave region of the electromagnetic spectrum is found below that of ultraviolet, visible and infrared and is classified as that between 300 and 300,000 MHz (Figure 1). From a chemistry perspective, microwaves are not of high enough energy to break chemical bonds: they can only make molecules rotate. This is very different from the more energetic ultraviolet radiation which can break bonds when it interacts with molecules.

Microwaves, like all electromagnetic radiation, move at the speed of light and consist of oscillating electric and magnetic fields (Figure 2). These components oscillate at right angles to each other and to the direction of propagation. There are two ways that microwaves can heat a sample, both involving the interaction of molecules in the sample with the electric field of the microwave irradiation.

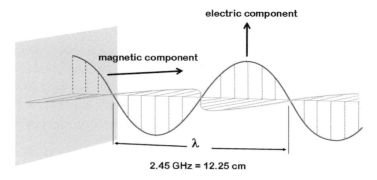

Figure 2 Microwave radiation consists of electric and magnetic fields.

If a molecule possesses a dipole moment, then when it is exposed to microwave radiation the dipole tries to align with the applied electric field. Since the electric field is oscillating, the dipoles constantly try to realign to follow this. At 2.45 GHz, molecules have time to align with the electric field, but not to follow the oscillating field exactly. This continual re-orientation of the molecules results in friction and thus heat. This heating method is termed dipolar polarization. If a molecule is charged, then the electric field component of the microwave irradiation moves the ions back and forth through the sample, also causing them to collide into each other. This movement again generates heat and is termed ionic conduction. The two mechanisms by which microwave energy leads to heating are shown in Figure 3.

dipolar polarization   **ionic conduction**

Figure 3   The two mechanisms by which microwave energy leads to heating.

The first reports of the use of microwave heating for organic chemistry appeared in 1986. Two research groups published results they had obtained in their laboratories using simple domestic microwave ovens. They found that the reactions they studied were completed much faster when they used the microwave oven than when they used conventional heating. However, there are some serious problems associated with using domestic microwave ovens for chemistry. First and foremost, they are not designed for containment of organic solvents and reagents and are therefore unsafe for this use. Also, when using an unmodified domestic microwave oven it is not possible to measure the temperature of a reaction accurately, and the microwave power is generally controlled by on-off cycles of the magnetron (pulsed irradiation).

More recently, scientific microwave equipment has become commercially available and is leading to increased popularity of microwave heating as a tool for preparative chemistry in the laboratory. As well as building reaction vessels able to withstand explosions inside the microwave cavity, temperature and pressure monitoring has been introduced as well as the capability of stirring reaction mixtures. Using small microwave units it is possible to run one reaction at a time and, with larger units, a number of reactions can be performed simultaneously, the samples being placed in sealed tubes and loaded onto a turntable.

Smaller microwave units are designed such that the cavity in which the sample is placed is the length of only one wave (termed a mode) and, as such, are called monomode apparatus. By placing the sample in the middle of the cavity it can be irradiated constantly with microwave energy (Figure 4). Using a monomode microwave unit, it is possible to heat samples as small as 0.2 mL very effectively. The upper volume limit of the monomode apparatus is determined by the size of the microwave cavity and is in the region of 100 mL. For larger samples, or for performing multiple reactions simultaneously, a larger microwave cavity is required. As the microwaves enter the cavity they move around and bounce off the walls. As they do so, they generate pockets of high

The cavity of a monomode microwave apparatus is designed for the length of only one mode

Figure 4   A monomode microwave unit.

energy and low energy as the moving waves either reinforce or cancel each other out. These microwave units are termed multimode (Figure 5). The microwave field in the microwave cavity is not uniform. Instead, there will be hot spots and cold spots, these corresponding to the pockets of high and low energy respectively. The field can be somewhat homogenized throughout the microwave cavity by using a device known as a mode stirrer. It is possible to perform reactions in open flasks up to a volume of 5 L using multimode apparatus. When using sealed vessels, the maximum capacity is approximately 1 L because of the safety issues associated with working at elevated temperatures and pressures. When multiple reactions are performed in a multimode unit, heating is fairly uniform as long as the reaction mixtures are quite similar, since as the samples move around they are large enough to absorb the microwave energy effectively.

For a multimode microwave unit, a power output of up to 1600 W is possible. For monomode apparatus the maximum power output is only 300–400 W. At first glance it may seem that the monomode apparatus is much less powerful than its multimode counterpart. However, the relative size of the cavity needs to

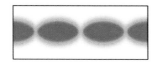

The cavity of a multimode microwave apparatus contains a number of modes

Figure 5   A multimode microwave unit.

be considered. Multimode microwaves have large cavities, and power is therefore dissipated over a large area. Monomode equipment has much smaller cavities, and the energy density is up to some 30–40 times higher than in the multimode apparatus.

Compared to using a hotplate to heat a reaction mixture, microwave irradiation is much more efficient and also greatly reduces the reaction time. The fact that heating occurs on a molecular level when using microwave irradiation, as opposed to relying on convection currents and thermal conductivity when a hotplate is used, explains why microwave reactions are so much faster. With microwave irradiation, since the energy is interacting with the molecules at a very fast rate, the molecules do not have time to relax and the heat generated can be, for short times, much greater than the overall recorded temperature of the bulk reaction mixture. In essence, there will be instantaneous localized superheating.

Consider a simple reaction with an activation energy of 200 kJ mol^{-1} performed at 150 °C. In order to get a 10-fold rate enhancement it would be necessary to increase the temperature by just 17 °C, and for a 1000-fold rate enhancement by 56 °C. Microwave energy provides instantaneous localized superheating to a much greater extent than this, which explains the enhanced reaction rates seen in chemical reactions when using microwave heating as compared to using a hotplate or any other conventional method.

The use of microwave heating is consistent with a number of the green chemistry principles. It is more energy efficient than conventional heating. Since it is often possible to obtain higher yields of a desired molecule using a microwave approach, there will be less waste and unused reagents. Also, since microwave heating is fast, there is often not enough time for products to decompose, so this makes the product purification cleaner and easier. Chemists have also used the inherent advantages of microwave heating to their advantage for developing cleaner alternatives to known reactions. Take for example the use of water as a solvent instead of organics like dichloromethane and benzene. Work has shown that water is an excellent solvent for organic chemistry, especially when combined with microwave heating. It is possible to heat water well above its boiling point in a sealed reaction vessel very safely and efficiently using microwaves. At these higher temperatures, water behaves more like an organic solvent. While most organic compounds are not soluble in water at room temperature, they can be in this higher temperature water, or at least partly so. This means that reactions can take place and, when the mixture cools down at the end, the product crystallizes out and is easily removed. As well as enabling a more environmentally friendly solvent to be used, it also makes purification easy. It is also possible to perform chemistry in water using open reaction vessels, simply under reflux. Again, microwave heating can be used to do this safely and efficiently.

Another advantage of using microwave heating is that it is often possible to reduce the amount of catalyst that is needed for a chemical reaction. Transition metal catalysts work very well under microwave irradiation. Catalytic cycles can turn around an astonishing number of times. In the experiment described here, the advantages of microwave heating in a metal-catalyzed reaction are clearly seen. This, coupled with the use of water as a solvent makes for fast, easy, clean chemistry.

Palladium-catalyzed reactions are some of the most widely used carbon–carbon bond forming reactions in organic synthesis today. One example of a palladium-catalyzed reaction is the coupling of aryl or alkenyl halides with alkenes to form more highly substituted alkenes. This reaction, which is now called the Heck-Mizoroki reaction, was first reported in the early 1970s by Mizoroki and Heck (Scheme 1). Traditional conditions use 1–2 mole % of a palladium catalyst (often bearing phosphine ligands), a base and an organic solvent. The temperature at which the coupling is performed as well as the time taken for the reaction to reach completion depend on the substrates used.

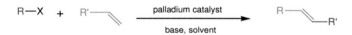

R = aryl, alkenyl R' = alkyl, aryl, alkenyl X = I, Br, OTf, Cl

Scheme 1

The active palladium catalyst is $Pd(0)L_n$ which is generated *in situ* via reduction of a Pd(II) complex (Scheme 2). The palladium then undergoes an oxidative addition, inserting into the C–X bond of the organo-halide component to form the Pd(II) complex **1**. Complex **1** reacts with the alkene component to form complex **2** via *syn*-addition to the double bond. Complex **2** rotates into complex **3**, which then undergoes a *syn* β-elimination, releasing the desired product and a hydrido-palladium (II) complex (**4**). Reductive elimination of HX from complex **4** regenerates the active form of the catalyst to restart the catalytic cycle.

While the Heck reaction is best suited for the preparation of trans-disubstituted alkenes from terminal alkenes, more highly substituted alkenes have been

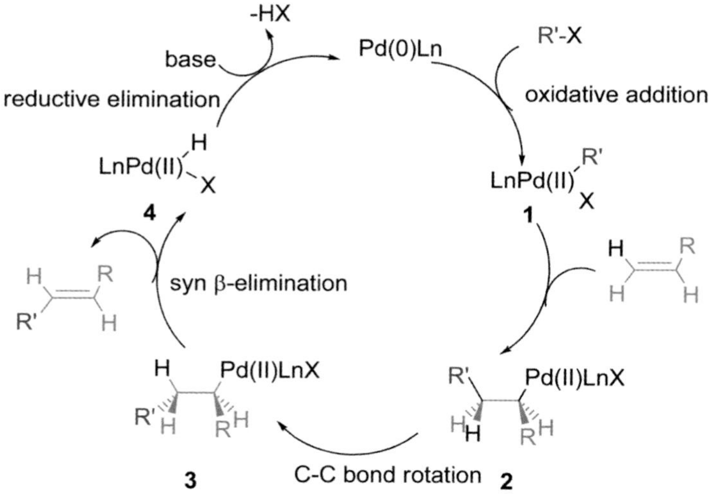

Scheme 2

synthesized using this methodology. The rate of the reaction decreases as the substitution on the alkene substrate increases, though the reaction is tolerant of a wide range of alkene substituents of varying electronic properties. The alkyl halide component is usually aryl or alkenyl in nature, though alkyl halides without β-hydrogens have been used.

While it is possible to use low levels of palladium catalysts without ligands for the Heck reaction in conjunction with N-methyl-2-pyrrolidinone (NMP) as a solvent, the procedure can be so slow as to become impractical. However, when using water as a solvent in conjunction with microwave heating it is possible to perform Heck reactions using as little as 500 ppb to 5 ppm of palladium salts as catalysts. The reactions are complete after about 5 min of heating. By using such low catalyst loadings it is often possible to negate the need for a costly metal extraction process at the end of the reaction. Also, since the product from the reaction is usually insoluble in water, product isolation at the end of the reaction is simplified. In this experiment, the coupling of styrene with either p-bromoanisole or p-bromotoluene is performed using a version of this low-catalyst loading protocol.

Some experiments using microwave heating

Preparation of 4-methoxystilbene

MW, 170 °C, 10 min

Pd stock solution
K$_2$CO$_3$, TBAB, H$_2$O

Apparatus	Glasschem 20 mL microwave glass reaction vessel with Teflon Screw cap for the CEM MARS5 multimode microwave or 10 mL reaction tubes with septum for the CEM Discover monomode microwave, magnetic stirrer, syringe (1 mL) or Eppendorf pipette, safety glasses, laboratory coat, protective gloves.
Chemicals	Styrene, tetra-*n*-butylammonium bromide (TBAB), bromoanisole, potassium carbonate, palladium stock solution (1000 ppm in water), distilled water.
Attention!	Safety glasses and protective gloves must be worn at all times.
Caution!	Because of their toxicity and volatility, care should be taken to avoid inhalation of bromoanisole and styrene. Tetra-*n*-butylammonium bromide (TBAB) and sodium carbonate are classified as irritants; avoid contact of their solutions with the skin. All reactions should be carried out in a well-ventilated hood. This reaction should not be attempted in a sealed reaction vessel without temperature control. The reaction solutions should cool to 50 °C before removing them from the oven.
Procedure	To a 10–20 mL microwave glass reaction tube are added *p*-bromoanisole (187 mg, 0.125 mL, 1.0 mmol), styrene (208 mg, 0.230 mL, 2.0 mmol), K$_2$CO$_3$ (511 mg, 3.7 mmol), tetra-*n*-butylammonium bromide (322 mg, 1.0 mmol), palladium stock solution (0.4 mL of a 1000 ppm solution in water [1]) and water (1.6 mL) to give a total volume of water of 2 mL and a total palladium concentration of 200 ppm. The reaction vessel is sealed with a Teflon screw cap, tightened with the preset microwave torque wrench (snap cap for the Discover), shaken and placed in the microwave cavity. The temperature thermocouple is connected to the control vessel which occupies position #1 in the

MARS5 microwave. (This is not needed for the Discover microwave.) The microwave is programmed using the standard heating profile with the initial microwave irradiation of 600 W (300 W for the Discover), a target temperature of 170 °C and a 10 min hold time. The temperature is ramped from room temperature to the desired temperature of 170 °C. Once this temperature is reached, the reaction mixture is held at this temperature for 10 min. The mixture is allowed to cool in the microwave for 20 min or until it is below 50 °C before the reaction vessel is opened. The contents are poured into an Erlenmeyer flask. Ethyl acetate (20 mL) is added to the Erlenmeyer flask and the solution is swirled until the entire white solid is dissolved. Water (20 mL) is added and the reaction mixture is transferred to a separatory funnel (60 mL). The separatory funnel is stoppered and repeatedly shaken, and the solution is vented. The layers are allowed to separate. The bottom aqueous layer is removed to an Erlenmeyer flask and the upper organic layer is collected in a separate Erlenmeyer flask. The aqueous layer is returned to the separatory funnel and extracted with ethyl acetate (10 mL) two more times. The top organic layer is collected, combined with the previous organic layers, and washed with saturated sodium chloride solution to remove any remaining water. The organic layer is then dried over $MgSO_4$, and the ethyl acetate is removed *in vacuo* leaving the crude product. The beige solid is crystallized from *n*-hexane. The shiny white crystals are collected by vacuum filtration and dried under vacuum.

Characterization data: m.p. 134–135 °C. ^1H NMR ($CDCl_3$): δ 7.46–7.53 (m, 4H), 7.37 (t, J = 7.6 Hz, 2H), 7.26 (t, J = 7.3 Hz, 1H), 7.10 (d, J = 16.3 Hz, 1H), 7.00 (d, J = 16.3 Hz, 1H), 6.93 (dd, J = 8.8 Hz, J = 1.9 Hz, 2H), 3.85 (s, 3H) ppm. ^{13}C NMR ($CDCl_3$): δ 159.4, 137.7, 130.2, 128.7, 128.3, 127.8, 127.2, 126.7, 126.3, 114.2, 55.3 ppm.

Preparation of 4-methylstilbene

Apparatus   Glasschem 20-mL microwave glass reaction vessel with Teflon screw cap for the CEM MARS5 multimode microwave or 10 mL reaction tubes with septum for the CEM Discover monomode microwave, magnetic stirrer, syringe (1 mL) or Eppendorf pipette, safety glasses, laboratory coat, protective gloves.

Chemicals   Tetra-*n*-butylammonium bromide (TBAB), *p*-bromotoluene, styrene, potassium carbonate, palladium stock solution (1000 ppm in water), distilled water.

Attention!   Safety glasses and protective gloves must be worn at all times.

Caution!   Because of its toxicity and volatility, care should be taken to avoid inhalation of *p*-bromotoluene, and styrene. Tetra-*n*-butylammonium bromide (TBAB) and sodium carbonate are classified as irritants; avoid contact of their solution with the skin. All reactions should be carried out in a well-ventilated hood. This reaction should not be attempted in a sealed reaction vessel without temperature control. The reaction solutions should cool to 50 °C before removing them from the oven.

Procedure   To a 10–20 mL microwave glass reaction tube are added *p*-bromotoluene (171 mg, 1.0 mmol), styrene (208 mg, 0.230 mL, 2.0 mmol), K_2CO_3 (511 mg, 3.7 mmol), tetra-*n*-butylammonium bromide (322 mg, 1.0 mmol), palladium stock solution (0.4 mL of a 1000 ppm solution in water [1]) and water (1.6 mL) to give a total volume of water of 2 mL and a total palladium concentration of 200 ppm. The reaction vessel is sealed with a Teflon screw cap, tightened with the preset microwave torque wrench (snap cap for the Discover), shaken and placed in the microwave cavity. The temperature thermocouple is connected to the control vessel which occupies position #1 in the MARS5 microwave. (This is not needed for the Discover microwave.) The microwave is programmed using the standard heating profile with the initial microwave irradiation of 600 W (300 W for the Discover), a target temperature of 170 °C and a 10 min hold time. The temperature is ramped from room temperature to the desired temperature of 170 °C. Once this temperature is reached, the reaction mixture is held at this temperature for 10 min. The mixture is allowed to cool in the microwave for 20 min or until it is below 50 °C before the reaction vessel is opened. The contents are poured into an Erlenmeyer flask. Ethyl acetate (20 mL) is added to the Erlenmeyer flask and the solution is swirled until the entire white solid is dissolved. Water (20 mL) is added and the reaction mixture is transferred to a separatory funnel (60 mL). The separatory funnel is stoppered and repeatedly shaken, and the solution is vented. The layers are allowed to separate. The bottom aqueous layer is removed to an Erlenmeyer flask and the upper organic layer is collected in a separate Erlenmeyer flask. The aqueous layer is returned to the separatory funnel and extracted with ethyl acetate (10 mL) two more times. The top organic layer is collected and combined with the previous organic layers, and washed with saturated sodium chloride solution to remove any remaining water. The organic layer is then dried over $MgSO_4$, and the ethyl acetate is removed *in vacuo* leaving the crude product. The beige solid is crystallized from ethanol. The shiny white crystals are collected by vacuum filtration and dried under vacuum.

Characterization data: m.p. 117–119 °C. ^{1}H NMR (CDCl$_3$): δ 7.52 (d, $J = 7.5$ Hz, 2H), 7.44 (d, $J = 7.9$ Hz, 2H), 7.37 (t, $J = 7.5$ Hz, 2H), 7.26–7.24 (m, 1H), 7.19 (d, $J = 7.7$ Hz, 2H), 7.10 (s, 2H), 2.38 (s, 3H) ppm. ^{13}C NMR (CDCl$_3$): δ 137.7, 137.6, 134.7, 129.5, 128.8, 127.8, 127.5, 126.6, 126.5, 21.3 ppm.

Waste Disposal Dispose of all waste in the appropriately marked containers in the fume hoods. Be sure to place organics only in containers labeled »organic waste«. Aqueous waste containing excess palladium salts should be placed in a designated »waste heavy metals« container in the laboratory. Dismantle and clean all glassware thoroughly with soap and water.

Note

1 The palladium loading is 0.4 mole%. One can substitute Pd(OAc)$_2$ (0.9 mg, 0.004 mmol) for the palladium stock solution. When using palladium acetate for laboratory classroom purposes we have found it convenient to premix the palladium acetate (16 mg) with the potassium carbonate (10 g). Students then only need to weigh out the amount of potassium carbonate required and do not need to add the palladium catalyst separately. The water solvent should also be increased to 2 mL.

18

Ionic Liquids as Benign Solvents for Sustainable Chemistry

Anders Riisager and Andreas Bösmann

Carbohydrates obtained from biomass via its major biopolymer constituents – cellulose, hemicellulose and lignin – are currently attracting considerable interest as readily available, relatively inexpensive and renewable feedstocks for the chemical industry. Traditional cellulose dissolution processes, including the cuprammonium and xanthate processes (employing ammoniacal copper and carbon disulfide solutions, respectively) are, however, often cumbersome and energy demanding because they require the use of relatively harsh conditions. Moreover, these processes cause serious environmental problems because the solvents cannot be recovered and reused. Similarly, direct transformation of underivatized carbohydrates to other organic products is quite challenging due to their low solubility in almost any solvent but water. The few solvents in which they will dissolve, such as DMF and DMSO, have highly undesirable environmental characteristics and are not compatible with many intended applications of carbohydrate-derived products.

Ionic liquids are a group of new organic salts that exist as liquids at a relatively low temperature ($< 100 °C$). Ionic liquids have tuneable physicochemical properties, usually negligible vapor pressures and, in general, thermal stability over a wide range of temperatures [1]. As a result, they have been used successfully to replace traditional solvents employed in a variety of synthetic and manufacturing processes, and have the potential to reduce the present reliance of industry on volatile organic compounds (VOCs). In addition to their role as alternative reaction and extraction media, certain ionic liquids have demonstrated to

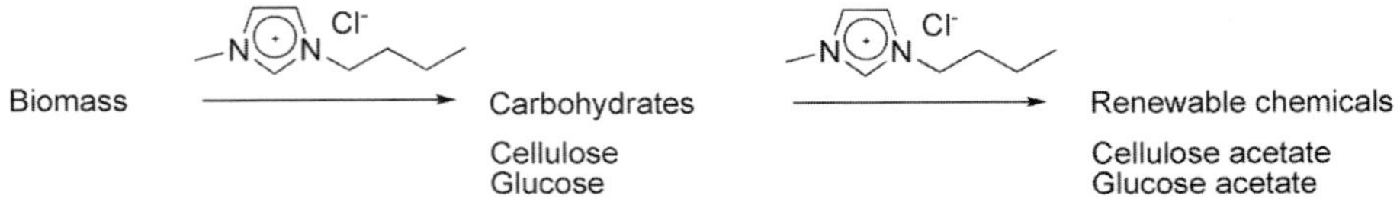

Biomass $\longrightarrow$ Carbohydrates $\longrightarrow$ Renewable chemicals
Cellulose Cellulose acetate
Glucose Glucose acetate

Scheme 1

dissolve polysaccharides and other biopolymers very efficiently, thus facilitating the development of greener reaction technologies [2]. Perhaps the most salient results in this regard have been those obtained with the common ionic liquid 1-*n*-butyl-3-methylimidazolium chloride, [BMIM]Cl, (see Scheme 1), which can dissolve cellulose [3], lignin [4, 5] and wood [6, 7] in relative high concentrations with no prior derivatization, and thus provide a new platform for the utilization of biomaterials as renewable feedstocks.

Some practical experiments illustrating the preparation and use of [BMIM]Cl

Preparation of 1-*n*-butyl-3-methylimidazolium chloride, [BMIM]Cl

Apparatus	500 mL three-necked round-bottomed flask, reflux condenser, magnetic stirrer, oil bath, T-shaped inert gas inlet, safety glasses, laboratory coat, protective gloves.
Chemicals	*N*-Methylimidazole, 1-chlorobutane, toluene, ethyl acetate.
Attention!	Safety glasses and protective gloves must be worn at all times!
Caution!	Because of their toxicity and volatility, care should be taken to avoid inhalation of the solvents and chemicals or contact of their solution with the skin. All reactions should be carried out in a well-ventilated hood.
Procedure	Preferably, *N*-methylimidazole is freshly distilled from KOH, toluene is freshly distilled from sodium and 1-chlorobutane is freshly distilled from CaH_2.

A 500 mL three-necked round-bottomed flask equipped with magnetic stirrer, reflux condenser, and T-shaped inert gas inlet is charged with a solution of *N*-methylimidazole (82.1 g, 1 mol) in toluene (100 mL). 1-Chlorobutane (111.1 g, 1.2 mol) is added at room temperature under inert gas atmosphere and the resulting solution is heated to reflux under stirring at ca. 110 °C for 24 h, after which it is placed in a freezer at ca. −20 °C for 12 h. The toluene is decanted and the remaining viscous oil/semi-solid is repeatedly recrystallized from ethyl acetate to yield a white crystalline solid, which is dried *in vacuo* to give [BMIM]Cl in approximately 80% yield.

Tip: If a yellowish coloration of the [BMIM]Cl can be accepted, the distillation of the chemicals can be omitted if they are of sufficient dryness.

The progress of the reaction can be followed by the formation of a biphasic system (bottom layer is [BMIM]Cl) from the initially monophasic solution. If required, the end of the reaction can be determined by mixing a drop of the bottom layer with 1-mL aqueous solution containing 1 wt% $CuSO_4$. An intensification of the blue color indicates residual *N*-methylimidazole.

Characterization data: m.p. 67 °C. ^1H NMR (D_2O): δ 0.95 (3H, t), 1.38 (2H, sextet), 1.80 (2H, q), 4.00 (3H, s), 4.30 (2H, t), 7.00 (1H, s), 7.15 (1H, s), 7.55 (1H, s) ppm. ^{13}C NMR (D_2O): δ 15.72, 34.21, 36.04, 38.69, 52.14, 124.16, 126.39, 141.16 ppm.

Explanation

Although crystalline at ambient conditions [BMIM]Cl is categorized as an ionic liquid as its melting point is below 100 °C. The ionic liquid is very hygroscopic and care must be taken to avoid water contamination, which will prevent the crystallization of the material. Its main use is as a precursor to a variety of other ionic liquids, namely ionic liquids of the type [BMIM][M_nCl_m], which are prepared from [BMIM]Cl by the addition of a metal chloride, e.g. $AlCl_3$. Other common ionic liquids incorporating the [BMIM]$^+$ cation are prepared by anion exchange, e.g. [BMIM][$(CF_3SO_2)_2N$]. [BMIM]Cl itself can be used to dissolve carbohydrates and biopolymers such as cellulose and silk in high concentrations [8, 9]. Toxicological tests have shown that [BMIM]Cl must be regarded as toxic, whereas this is not the case for [EMIM]Cl, which may be used alternatively. The melting point of [EMIM]Cl is about the same as that of [BMIM]Cl.

Dissolution and precipitation of Cellulose in [BMIM]Cl

Apparatus   Test tubes with screw cap or 25 mL round-bottomed flask with stopper, safety glasses, laboratory coat, protective gloves.

Chemicals   [BMIM]Cl, absolute ethanol, cellulose linters or cellulose filter paper.

Attention!   Safety glasses and protective gloves must be worn at all times.

Procedure   Cellulose linters or cellulose filter paper is dried at 80 °C under vacuum for 4 h.

A solution of 10 wt% cellulose in [BMIM]Cl is prepared by melting ionic liquid (9.0 g) at 90 °C and adding cellulose (1.0 g). The resulting mixture is stirred at this temperature until a clear, amber colored solution is obtained (this may take up to 12 h).

To regenerate the cellulose, 2.0 mL of the hot cellulose solution is added drop by drop into 30 mL of hot (65–70 °C) ethanol under vigorous stirring. After the droplets have dissolved, a suspension of cellulose fibers in ethanol is observed. The suspension is filtered and the fibers are washed with 3×5 mL ethanol and dried *in vacuo*. Yield of the regenerated cellulose is quantitative.

The combined ethanolic solutions are evaporated with a rotary evaporator at 60 °C and 100 mbar. The remaining [BMIM]Cl is then dried under high vacuum at 80 °C for 8 h and can subsequently be reused.

Characterization data: ^{13}C NMR (DMSO-d$_6$, 90 °C): δ 103.2, 74.7, 75.8, 80.2, 76.2, 61.1 ppm.

Explanation

There are very few solvents that can dissolve cellulose without derivatizing it, for example cuprammonium hydroxide, *N*-methylmorpholine-*N*-oxide or a mixture of *N,N*-dimethylacetamide and LiCl. A recent development is the use of ionic liquids with anions like chloride that are capable of disrupting the extensive hydrogen bonding network of the cellulose. The dissolution of cellulose is a prerequisite for characterization, shaping (e.g. fiber spinning) and further processing using traditional organic transformations in homogeneous solution.

Ultrasound-mediated acetylation of glucose in [BMIM]Cl

Apparatus   25 mL two-necked flask, reflux condenser, septum, syringe (10 mL) and thermostattically controlled ultrasonic cleaning bath, safety glasses, laboratory coat, protective gloves.

Chemicals	[BMIM]Cl, glucose, petroleum ether, acetic anhydride, ethyl acetate, Na_2CO_3.
Attention!	Safety glasses and protective gloves must be worn at all times.
Procedure	Glucose (1.8 g, 10 mmol) is dissolved in 5.0 g [BMIM]Cl at 80 °C under a dry atmosphere. Acetic anhydride (5.62 g, 55 mmol) is added carefully via a syringe. The mixture is then sonicated for 15 min and the resulting acetyl ester is extracted with 3 × 30 mL petroleum ether:ethyl acetate (1:5). The combined organic phase is dried over Na_2CO_3 and the solvents removed on a rotary evaporator at 50 °C and 100 mbar. Yield 3.42 g (91 %). The ionic liquid is regenerated by removing volatiles under high vacuum at 80 °C for 2 h.

Characterization data: m.p. 131 °C.

Explanation

Acetylation of alcohols, one of the most frequent reactions in carbohydrate chemistry, is used to protect the hydroxyl group during a multi-step synthesis. The reaction is usually carried out using acetic anhydride and a variety of catalysts in different solvents [10]. Ultrasound-mediated *O*-acetylation of alcohols in the ionic liquid [BMIM]Cl gives high yields of the acetylated compounds in relatively short time without the need of a catalyst, compared to the yields obtained with reactions carried out under »silent« conditions or in conventional molecular solvents [11]. Moreover, the acetylation is stereospecific with retention of configuration when optically active starting material (such as *D*-(+)-glucose) is employed. It is assumed that the Lewis/Brønsted acidity of the imidazolium cation is responsible for the ability of the ionic liquid to promote the reaction.

Acetylation of Cellulose in [BMIM]Cl

Apparatus	25 mL two-necked flask, reflux condenser, septum and syringe (5 mL), safety glasses, laboratory coat, protective gloves.
Chemicals	10 wt% solution of cellulose in [BMIM]Cl, acetyl chloride, absolute ethanol.

Attention!   Safety glasses and protective gloves must be worn at all times.

Procedure   Acetyl chloride (1.09 mL, 15.3 mmol) is carefully added via a syringe to 5.0 g of a 10% solution of cellulose (i.e. 5 mol acetyl chloride per mol anhydroglucose unit) in [BMIM]Cl at 80 °C. The temperature is kept at 80 °C for another 2 h. The product is then isolated by precipitation in 100 mL ethanol, washing with 2 × 50 mL ethanol and finally drying under vacuum at 60 °C. Yield 0.75 g (86%).

Characterization data: ^{13}C NMR (DMSO-d$_6$): δ 168.9-170.2 (C=O), 99.2, 75.9, 71.3 and 62.1 (cellulose backbone) ppm. IR (KBr, cm^{-1}) $\bar{v}$ 2890 (CH), 1750 (C=O). DS$_{acetate}$ = 3.0 (determined by means of ^{1}H NMR).

Explanation

Cellulose is the most abundant organic polymer in nature. It is a very uniform macromolecule that is reactive, biocompatible, biodegradable, chiral and stereo regular. As a renewable polymer, it must be considered as a future raw material for polymers. Cellulose esters are used for the production of artificial fibers, varnishes, photographic materials, plastics and other applications. In the O-acetylation of cellulose in [BMIM]Cl, very high values of DS (degree of substitution) can be achieved [12]. This a result of the homogeneous nature of the reaction, which is made possible by the relatively high solubility of cellulose in the ionic liquid.

References

1 P. Wasserscheid, T. Welton (Eds.), *Ionic Liquids in Synthesis 2nd ed.*, Wiley-VCH, Weinheim, **2007**.
2 Q. Liu, M.H.A. Janssen, F. van Rantwijk, R.A. Sheldon, *Green Chem.* **2005**, *7*, 39.
3 R.P. Swatloski, S.K. Spear, J.D. Holbrey, R.D. Rogers, *J. Am. Chem. Soc.* **2002**, *124*, 4974.
4 H. Xie, A. King, I. Kilpelainen, M. Granstrom, D.S. Argyropoulos, *Biomacromolecules* **2007**, *8*, 3740.
5 Y. Pu, N. Jiang, A.J. Ragauskas, *J. Wood Chem. Technol.* **2007**, *27*, 23.
6 D.A. Fort, R.C. Remsing, R.P. Swatloski, P. Moyna, G. Moyna, R.D. Rogers, *Green Chem.* **2007**, *9*, 63.
7 I. Kilpeläinen, H. Xie, A. King, M. Granstrom, S. Heikkinen, D.S. Argyropoulos, *J. Agric. Food Chem.* **2007**, *55*, 9142.
8 J.S. Moulthrop, R.P. Swatloski, G. Moyna, R.D. Rogers, *Chem. Commun.* **2005**, 1557.
9 O.A. El Seoud, A. Koschella, L.C. Fidale, S. Dorn, T. Heinze, *Biomacromolecules* **2007**, *8*, 2629.
10 T.W. Greene, P.G.M. Wuts, *Protective Groups in Organic Synthesis 2nd ed.*, Wiley, New York, **1991**.
11 A.R. Gholap, K. Venkatesan, T. Daniel, R.J. Lahoti, K.V. Srinivasan, *Green Chem.* **2003**, *5*, 693.
12 S. Barthel, T. Heinze, *Green Chem.* **2006**, *8*, 301.

19

Olefin Self-Cross Metathesis in an Ionic Liquid

Megumi Fujita

Since the late 1990s, the number of studies of room temperature ionic liquids (RTILs) as new reaction media has been growing exponentially [1, 2]. RTILs have drawn attention in the context of green chemistry as a nonvolatile and reusable alternative to organic solvents. RTILs that are widely used today are good solvents for a wide range of both inorganic and organic substances, and the solubility can be tailored by changing substituents on the component ions.

The cations of RTILs are generally bulky ammonium or phosphonium ions with low charge densities. The 1,3-dialkylimidazolium cations are the most widely investigated. The commonly used anions include $[PF_6]^-$, $[BF_4]^-$, $[CF_3SO_3]^-$ and $[(CF_3SO_2)_2N]^-$.

When RTIL is used as a solvent for transition metal-catalyzed organic reactions, one of the main advantages is the recyclability of the catalyst. Organic products can be isolated from the RTIL–catalyst mixture by distillation or extraction with another solvent, and the catalyst solution can be used repeatedly.

The purpose of the experiment described below is to examine (a) the efficiency of olefin self-cross metathesis catalyzed by one type of Grubbs-Hoveyda second-generation catalyst (1) in one of the most common ionic liquids 1-butyl-3-methylimidazolium hexafluorophosphate ([BMIM]PF_6, 2), and (b) the recyclability of the catalyst after the first reaction. Olefin metathesis is a powerful method for creating new carbon–carbon double bonds, and generations of efficient catalysts have been developed. However, these catalysts require chromatography separation from the organic products when reaction is carried out in traditional organic solvents, and the catalysts are technically nonrecyclable. This is a severe drawback in industrial applications. If an RTIL is used as a reaction medium in which the metathesis catalyst has an excellent solubility, the products can be separated from the medium, while the RTIL retains the catalyst.

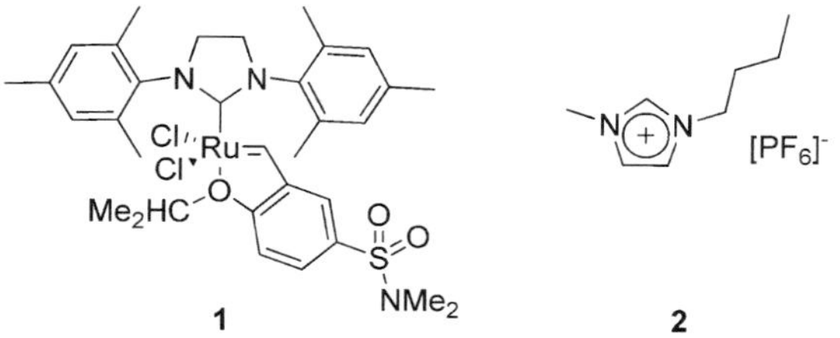

Scheme 1

This experiment is adopted from published work [3–5], especially by Ding et al. [3]. This experiment is optimized for four 3-h laboratory sessions over two weeks. A simple nitrogen line equipped with an oil bubbler at the end and with several T-branches, each with tubing ending with a syringe needle outlet, may be used to accommodate multiple student experiments. It is recommended to carry out the initial air purging by nitrogen bubbling one reaction vessel at a time while closing other outlets (including the bubbler) to concentrate the nitrogen flow.

Some practical demonstrations of olefin self-cross metathesis in ionic liquid media

Catalysis reaction (first cycle)

Apparatus   Glass vial (20 mL capacity), septum, flea-size magnetic stirbar, magnetic stirrer, N_2 line with N_2 outlets equipped with syringe needles, microsyringe (250 or 500 µL), pipettes, 50-mL round-bottomed flask, thin-layer chromatography (TLC) chamber (beaker and watch glass), small chromatography column, test tubes for column fractions, NMR tube, UV lamp for TLC visualization, safety glasses, laboratory coat, protective gloves.

Chemicals   1,3-Bis(2,4,6-trimethylphenyl)-4,5-dihydroimidazol-2-ylidene[2-(i-propoxy)-5-(N,N-dimethylaminosulfonyl)-phenyl]methyleneruthenium(II) dichloride (**1**, Strem Chemicals), 1-butyl-3-methylimidazolium hexafluorophosphate ([BMIM]PF$_6$, **2**, Sigma-Aldrich), petroleum ether, styrene, silica TLC plate with fluorescent indicator, silica, DMSO-d$_6$.

Attention!   Safety glasses and protective gloves must be worn at all times.

Caution!   Avoid skin contact with [BMIM]PF$_6$ (irritant), styrene (flammable, irritant and carcinogen), petroleum ether (flammable, target organs: central nervous system, lungs, heart, liver, ears). All reactions should be carried out in a well-ventilated hood.

Procedure   The Ru catalyst (0.0183 g, 0.025 mmol) is weighed in a 20-mL glass vial. A flea-size stirbar and 2 mL of [BMIM]PF$_6$ are added and the vial is stoppered with a septum. The air is purged by bubbling N$_2$ through the gently stirred mixture via a syringe needle, while placing another vent syringe needle on the septum. After purging air for at least 10 min, the N$_2$ inlet needle is lifted above the liquid and the vent needle removed to keep the mixture under N$_2$. To the stirred solution, styrene (0.115 mL, 1 mmol) is added via a microsyringe. The mixture is stirred at room temperature until the next laboratory period (within two days).

Under the fume hood, the organic substances are extracted four times with petroleum ether (3 mL aliquots). Use a pipette to mix the two layers by bubbling into the mixture several times; let the layers separate and then draw off the top layer carefully. The extracts are collected in a preweighed round-bottomed flask. A TLC (silica gel plate, petroleum ether as an eluent) is carried out with starting styrene as a reference. The volatiles are removed by rotary evaporator. The mass of the residue (crude product) is recorded. The ionic liquid mixture containing the catalyst is dried under vacuum for at least 30 min to be used in the second cycle.

The product *E*-stilbene is purified by silica gel column chromatography using petroleum ether as an eluent. The yield of the product is recorded and the product is identified by ^{1}H NMR. The unreacted stilbene may or may not be observed as it may have been lost during the evaporation process.

Characterization data:

E-Stilbene: m.p. 122–125 °C. ^{1}H NMR: δ 7.03 (s, 2H), 7.10 (t, 2H), 7.25 (>t<, 4H), 7.41 (d, 4H) ppm.

Z-Stilbene: m.p. 5–6 °C. ^{1}H NMR: δ 6.57 (s, 2H), 7.18 (overlapped, 10H) ppm.

Styrene (unreacted reactant): b.p. 145°C. ^{1}H NMR: δ 5.22 (d, 1H), 5.70 (d, 1H), 6.70 (dd, 1H), 7.22 (t, 1H), , 7.30 (>t<, 2H) 7.39 (d, 2H) ppm.

Catalyst/solvent recyclability (second cycle)

Procedure   The ionic liquid–catalyst mixture from the first cycle is placed under N_2 in the same manner as in the first cycle. Styrene (0.115 mL) is added and the mixture is stirred for the same reaction time.

Extraction and purification procedures are carried out in the same manner as in the first cycle. Use TLC to roughly assess the reaction progress in the second round. Compare and report the product purity and yield between the first and second cycles.

Optional mini experiment: solubility tests

Procedure   After extraction of organic product and removal of volatiles at the end of the second cycle, the mixture of [BMIM]PF$_6$ and catalyst may be used to test for (a) miscibility of the ionic liquid with several solvents and (b) extraction of the catalyst into the solvent.

To a small test tube, 5–6 drops of the colored ionic liquid–catalyst mixture are added. A solvent (about 0.5 mL) is added and the mixture is shaken. Record the qualitative observations of miscibility and the extent of catalyst »leaching«. Suggested solvents to test are water, methanol, ethanol, acetone, ethyl acetate, tetrahydrofuran, diethyl ether, acetonitrile, dichloromethane, toluene and hexane.

Waste Disposal   All liquid wastes are discarded in a designated liquid waste container. The used silica is collected in a solid waste container.

References

1 V.I. Pâvulescu, C. Hardacre, *Chem. Rev.* **2007**, *107*, 2615.
2 T. Welton, *Coord. Chem. Rev.* **2004**, *248*, 2459.
3 X. Ding, X. Lv, B. Hui, Z. Chen, M. Xiao, B. Guo, W. Tang, *Tetrahedron Lett.* **2006**, *47*, 2921.
4 D.B.G. Williams, M. Ajam, A. Ranwell, *Organometallics* **2006**, *25*, 3088.
5 R.C. Buijsman, E. van Vuuren, J.G. Sterrenburg, *Org. Lett.* **2001**, *3*, 3785.

20

Experiments with Ionic Liquids

Gordon W. Driver and Keith E. Johnson

Over the last 20 years the field of ionic liquids (IL) chemistry has expanded enormously. Before this, the chemistry of simple liquid salts such as the alkali metal halides, with melting points of several hundred degrees, was well established and, of course, well utilized in the industrial preparation of iron, aluminum, and many reactive metals. The newer ionic liquids consist of organic cations, often quaternary nitrogen or equivalent atoms, and inorganic or simple organic anions. Examples of these cations are dialkylimidazolium, alkylpyridinium and trialkylsulfonium. Anions include halides, carboxylates and haloaluminates.

The attraction of these newer liquid salts is that many of them are usable at ambient temperatures, most being usable at 100 °C. Those with relatively good conductivities (i.e. those containing few ion pairs) show low volatility. With the appropriate choice of ions they are quite stable to oxidation and reduction as expressed in large electrochemical windows. The versatility of the systems can be controlled by the choice of cations and/or anions, including the derivatization of the ion building blocks.

Chemistry with these systems is readily related to that of certain high-temperature systems, but also to various well-known nonaqueous solvents. Acidity in ionic liquids tends to follow the gas phase properties of the molecules involved rather than the aqueous acid/base behavior with which we are all familiar.

In this text we shall discuss two reactions: (a) the cleavage of ethers using liquid 3-methylimidazolium dibromohydrogenate(I), [Hmim][HBr$_2$], an acid stronger than aqueous HBr, and (b) the addition of hydride ions to a ketone dissolved in triethylsulfonium hexachlorobromodialuminate, [Et$_3$S][Al$_2$Cl$_6$Br], to form an organic alkoxide – the type of reaction that would proceed to an alcohol in an aqueous medium.

Three practical demonstrations of reactions in ionic liquids

The cleavage of anisole in 3-methylimidazolium dibromohydrogenate(I)

The following experiment demonstrates the nature and utility of a Brønsted acid in a nonaqueous »ionic« solvent, namely 3-methylimidazolium dibromohydrogenate(I) or [Hmim][HBr$_2$] [1, 2]. This ionic liquid was specifically designed for the task of cleaving stable ethers efficiently at ambient temperatures with a view to an environmentally friendly and economic process for the cleavage of polymeric ether linkages found in lignocellulosic biomass feedstocks.

The solvent is formed by the neutralization of 1-methylimidazole with hydrogen bromide. In a 1:1 mole ratio, the nonacidic homogeneous catalyst [Hmim]Br is produced, which can be observed as either a solid salt (m.p. 75–80 °C) or a supercooled liquid. The addition of a second equivalent of hydrogen bromide provides the 1:2 system where a tuning of the anion initially leads to the formation of [HBr$_2$]$^-$. The bridged hydrogenate(I) species then disproportionates to form an equilibrium mixture of anionic species including Br$^-$, [HBr$_2$]$^-$, and [H$_2$Br$_3$]$^-$, according to ^{1}H NMR analyses. The two protic anions, dibromohydrogenate(I) and tribromodihydrogenate(I), are observed as the acidic species and are governed by the following equilibrium:

$$\text{Br}^- + [\text{H}_2\text{Br}_3]^- \rightleftharpoons 2[\text{HBr}_2]^-$$

Variable temperature (VT) ^{1}H NMR experiments of the neat liquid reveal a time-averaged single anionic proton signal (δ 5.53 ppm, relative to internal TMS) that moves upfield at increased temperatures indicating a shift of the equilibrium to the left. During the course of an ether reaction, the signal shifts downfield to higher ppm values as the concentration of bromide increases concomitant with the consumption of [HBr$_2$]$^-$. The nitrogenic proton of Hmim$^+$ was not labile to any observable extent according to VT ^{1}H NMR experiments. The N$^+$-H proton signal (δ 11.76 ppm relative to internal TMS) gave rise to a 1:1:1 triplet as a result of quadrupole splitting ($J(^{14}$N-^{1}H) = 62.5 Hz). Increased resolution and sharpening of the triplet with increases in temperature indicate that the N$^+$-H proton is stable and does not participate as an acid in the formation of oxonium species during the course of the ether cleavage reaction.

Apparatus   Assorted round-bottomed and pear-shaped flasks, activated molecular sieves, a double manifold/Schlenk line, distillation apparatus, magnetic stirring plate, magnetic stir bars, heating mantle and rheostat, rubber septa, locking plastic fasteners, syringe needles, airless syringe, a sealed capillary containing DMSO-d_6, an NMR sample tube, Teflon tubing, safety glasses, laboratory coat, protective gloves.

Chemicals

Anisole (Aldrich, 99%), 1-methylimidazole (Aldrich, 99%), anhydrous hydrogen bromide (Aldrich, ≥ 99%), crushed dry KOH, sodium metal, dry magnesium sulfate.

Caution!

Perform all manipulations/experiments in a fume hood. *Slowly* admit the first equivalent of $HBr_{(g)}$ into the reaction vial containing 1-methylimidazole (addition of the second equivalent is much less exothermic). As the neutralization reaction is extremely exothermic, periodically place the vial in an ice bath to prevent excessive heat build-up and subsequent decomposition of the base. Drying of anisole using sodium metal produces highly flammable $H_{2\,(g)}$ which must be vented in a fume hood. Sodium metal causes burns and reacts violently with water. The final product [Hmim][HBr$_2$] loses $HBr_{(g)}$ if not properly sealed. Material Safety Data Sheets (MSDS) should always be consulted when available.

Attention!

Safety glasses and protective gloves must be worn at all times.

Procedure

1. Purification of 1-methylimidazole: Pour 1-methylimidazole (approximately 15 mL) over crushed dry KOH (~1 g) contained in a clean and dry aluminum foil-covered, 25 mL round-bottomed flask and leave undisturbed for 24 h. Into a clean and dry aluminum foil-covered 25 mL round-bottomed flask containing activated 3-Å molecular sieves (~ 0.5 g) pour the predried 1-methylimidazole, taking care not to include any KOH. Once the distillation unit is set up (refer to Figure 1), purge with a cycle of vacuum/dry argon three times. With mild heating (~50 °C), twice distill the 1-methylimidazole under a blanket of argon at reduced pressure (~0.05 mm Hg), collecting each time on freshly activated 3 Å molecular sieves (~0.5 g). Store the final purified fraction (~8 mL) over sieves in an argon-purged 25 mL two necked round-bottomed flask (fitted with a purge valve and a rubber septum) in a refrigerator and keep covered with aluminum foil.

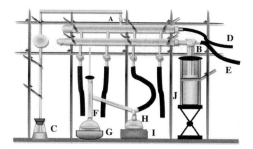

Figure 1 (A) double manifold/Schlenk line. (B) cold trap. (C) Hg bubbler. (D) dry $Ar_{(g)}$ line. (E) vacuum line. (F) distillation flask with condenser and thermometer. (G) heating mantle. (H) collection flask. (I) ice-water bath. (J) dewar containing $N_{2\,(l)}$.

2. Purification of anisole: Into a clean and dry 25 mL pear-shaped flask, pour anisole (15 mL). Carefully admit small pieces of cleaned sodium metal (~1 g total) to the flask and cap using a septum containing a small-bore needle to vent off any gas produced. (Metallic sodium is usually stored in paraffin oil which must be rinsed off with petroleum ether prior to use.)

Once all traces of water have been removed (i.e. formation of bubbles of $H_{2(g)}$ is no longer observable) attach the flask to a distillation apparatus. Twice distill the dried ether and store in a capped vial containing dry magnesium sulfate.

3. Preparation of [Hmim][HBr₂]: (1) To a preweighed septum-capped 20 mL glass vial (use a locking plastic fastener to tightly secure the inverted septum to the vial) admit the required quantity of purified 1-methylimidazole (~6 g) via an airless syringe. Using the set-up illustrated in Figure 2, *slowly* admit the first molar equivalent (~ 6 g) of anhydrous $HBr_{(g)}$. Cool the reaction vial periodically using an ice water bath. Observation of discoloration (slight yellowing) is indicative of excessive heat build-up causing decomposition. Repeatedly weigh the vial until the required weight of $HBr_{(g)}$ has been delivered. At this point, the 1:1 salt, [Hmim]Br, should appear as a highly viscous »supercooled« liquid. Intense cooling during the reaction may induce nucleation, in which case a white solid will be observed. (2) To the vial containing [Hmim]Br add a second molar equivalent of $HBr_{(g)}$. Protonation of bromide is much less exothermic than that of 1-methylimidazole, and this second equivalent of $HBr_{(g)}$ can be admitted more rapidly, although heat build-up must be monitored at all times. Repeatedly weigh the vial until the required

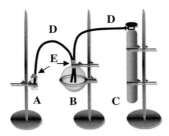

Figure 2 (A) septum-capped glass vial containing 1-methyl-imidazole. (B) gas trap. (C) lecture bottle of anhydrous HBr$_{(g)}$. (D) Teflon tubing. (E) rubber septa.

weight of HBr$_{(g)}$ has been delivered. The [Hmim][HBr$_2$] solvent should appear as a colorless (or possibly slightly yellow) highly fluid liquid.

4. The cleavage of anisole: To a clean and dry preweighed 25 mL round-bottomed flask containing a magnetic stir bar admit the quantity of [Hmim][HBr$_2$] prepared in Step 3 (based on the recommended ~6 g of 1-methylimidazole used in Step 3, approximately 10 mL of solvent will be available for use). With stirring, add a half molar equivalent of purified anisole (~4 g or 36.8 mmol) to the solvent, and cap the flask. The reaction can be monitored by sampling the reaction mixture and running $^{13}C\{^1H\}$ NMR. The sample should be run »neat« with a sealed capillary of DMSO-d_6 placed inside the NMR sample tube to facilitate field lock. Once signals due to the reactant are no longer observed, the reaction is complete to 95% (refer to Table 1 below). Phenol can be distilled directly from the reaction pot using the double manifold and mild heating. Note that unreacted quantities of HBr$_{(g)}$ and MeBr$_{(g)}$ produced are present after the reaction has completed and can be removed using purge cycles prior to distillation.

Characterization data: It is worthwhile noting that literature values of chemical shifts should not be used to verify products if they are referenced from a different solvent system. Instead, chemical shifts of authentic samples should be measured since solvent shifts of 5 ppm are possible. Literature values have been included in Table 1 to demonstrate this point.

Table 1 $^{13}C\{^1H\}$ NMR resonances of the ether and cleavage product. Literature values are included for comparison (IL = ionic liquid).

	δ (ppm, reported relative to TMS)				
Ether	Ether in IL	Product	Product in IL	Authentic sample in IL	Literature [3]
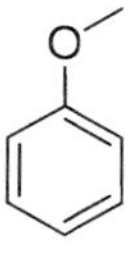	157.57	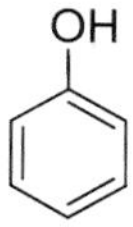	154.71	154.59	155.6
	128.36		128.43	128.51	130.5
	118.86		119.42	119.40	120.8
	112.58		114.24	114.33	116.1
	54.58				

Waste Disposal Sodium metal used in the drying of anisole can be destroyed using cold methanol. The ionic liquid can be disposed of with other laboratory halogenated waste. Unused portions of 1-methylimidazole and anisole should be properly labelled and secured in appropriate sealed vessels and stored.

Explanation

Anisole is cleaved to phenol and methyl bromide with ease using [Hmim][HBr$_2$] *via* an S$_N$2 reaction pathway. The protic oxonium cation (conjugate acid of the ether) intermediate forms as a result of proton transfer from [HBr$_2$]$^-$, as illustrated in Scheme 1.

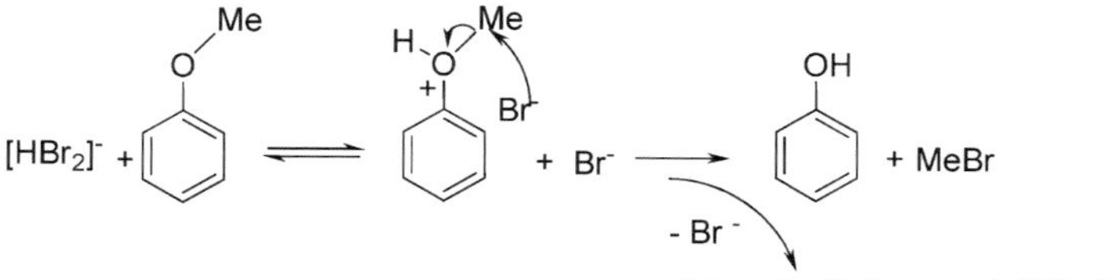

Scheme 1

   The oxonium species is short lived since activation is followed by rapid displacement of the alkyl group by Br$^-$. As [HBr$_2$]$^-$ is consumed in the reaction, a stoichiometric quantity of Br$^-$ is released and reacts with [H$_2$Br$_3$]$^-$ to produce more [HBr$_2$]$^-$. This behavior has been recorded using ^{1}H NMR where the time-averaged signal of the anionic mixture shifts to higher frequencies from the

$[H_2Br_3]^-$ limit of -3.69 ppm (vs. TMS) [4] towards the $[HBr_2]^-$ limit at δ 10.2 ppm (vs. TMS) [5]. After first use, recycling is possible with a »top up« of the $HBr_{(g)}$ content to restore the original 1:2 composition of $[Hmim][HBr_2]$.

The $-C-O-C-$ linkage of anisole is energetically very stable, making this ether, as well its analogs, generally difficult to cleave. A literature survey revealed that anisole cleaves in molecular solvents under harsher conditions. For example, solutions of Brønsted acids in various molecular solvents cleave anisole but at higher temperatures over longer periods, and require product workup while providing poorer yields [6–8]. Aqueous HBr solutions (48%, 8.68 mol L^{-1}) contain a higher concentration of acid than in $[Hmim][HBr_2]$ (7.17 mol L^{-1}), yet comparable reaction times are only achieved in the aqueous system through the use of a catalyst [9]. Lewis acid mixtures also effect cleavage of stable ethers but suffer from undesired side reactions and require higher reaction temperatures, increased molar equivalents of reagent, and quenching with aqueous workup to destroy the Lewis adduct in order to isolate the product [6, 10–12].

Reduction of benzophenone in triethylsulfonium hexachlorobromodialuminate

The following advanced experiment demonstrates the nature and utility of Lewis-acidic haloaluminate ionic liquids in the generation of a stabilized aroxide carbanion *via* addition of hydride to an aryl ketone. Lewis type ionic liquids have typically been utilized as catalytic Friedel–Crafts alkylation/acylation media and are known to be governed by various equilibria across the Lewis-basic, neutral and acidic compositions (Scheme 2).

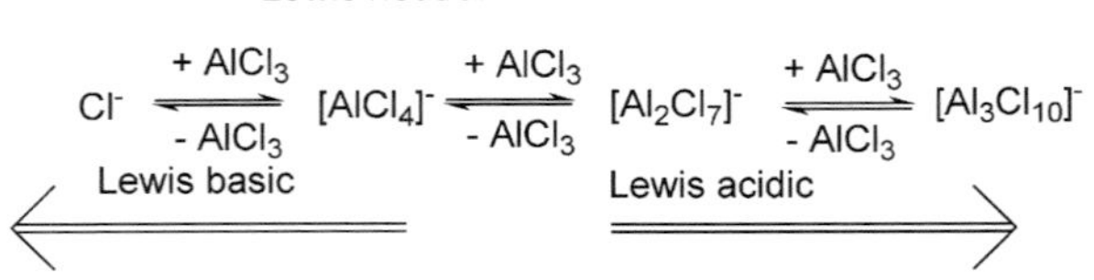

Scheme 2

Once within the > 0.50 mole fraction, χ, composition range, quantities of the homogeneous catalyst $[Al_2Cl_7]^-$ form, and at mole fractions of ≥ 0.67, quantities of $[Al_3Cl_{10}]^-$ begin to form, although in the latter composition range, melting points are known to increase causing freezing at room temperature. Addition of a Lewis base to the Lewis-acidic compositions of $0.50 < \chi \le 67$ causes $[Al_2Cl_7]^-$ to be neutralized and form $2[AlCl_4]^-$, (when the base is Cl^-). When the base is hydride, a hydridic species forms instead, according to [13]:

$$H^- + [Al_2Cl_7]^- \rightarrow [AlCl_4]^- + [AlCl_3H]^-$$

Additionally, $[AlCl_3H]^-$ can be generated *in situ* as a result of substrate hydride extraction in Lewis-acidic compositions and is known to be the active agent in the cracking of alkanes [14].

UV-Vis active naphthoquinones can be used for determination of $[AlCl_3H]^-$ colorimetrically since residual protic species, i.e. water, can consume quantities of hydride and interfere with stoichiometric formation of trichlorohydridoaluminate [15]. The reaction of 2,3-dichloro-1,4-naphthoquinone to form the aroxide, 2,3-dichloro-4-hydroxy-1-naphthoxide, is illustrated in Scheme 3 below.

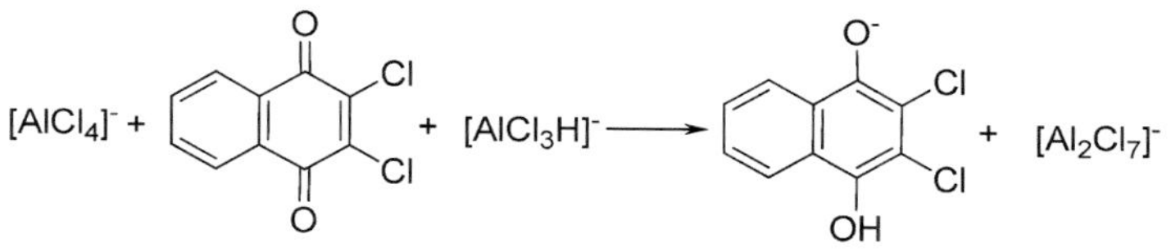

Scheme 3

It is noteworthy that, after the reduction proceeds, the homogeneous catalyst $[Al_2Cl_7]^-$ is regenerated. This reaction alternatively provides a route for the determination of impurity water with ionic liquids that may be sensitive to more conventional analyses, such as Karl Fischer coulometry.

Apparatus   Reflux condenser, assorted round-bottomed flasks, a double manifold/Schlenk line, medium-pore filter frit, an inert glove box, magnetic stirring plate/magnetic stir bars, heating mantle and rheostat, a sealed capillary containing DMSO-d_6, a NMR sample tube, Teflon tubing, safety glasses, laboratory coat, protective gloves.

Chemicals   Diethyl sulfide (Fluka, puriss. $\ge$ 99.0%), ethyl bromide (Aldrich, 99%), aluminum trichloride (Sigma-Aldrich, *Reagent Plus*, 99%), powdered calcium dihydride, benzophenone, freshly distilled acetonitrile, dry ethanol, dry diethyl ether.

Caution!
Perform all manipulations in a well-ventilated fume hood or inert glove box. Diethyl sulfide is extremely flammable and exerts a significant vapor pressure at ambient temperatures. Aluminum trichloride reacts violently with water. Reaction of $AlCl_3$ with $[Et_3S]Br$ is extremely exothermic: addition of the Lewis acid should proceed *very slowly* to avoid over-heated conditions and yellowing of the liquid formed! Liquid haloaluminates are water sensitive; keep the ionic liquid dry to avoid aluminum oxide formation and generation of HBr/HCl gases. Material Safety Data Sheets (MSDS) *should always* be consulted when available.

Attention!
Safety glasses and protective gloves must be worn at all times.

Procedure
1. Preparation of $[Et_3S][Al_2Cl_6Br]$: Treat chilled diethyl sulfide (~2.3 g, 0.025 mol) dissolved in freshly distilled, chilled acetonitrile (12 mL) with ethyl bromide (~3.9–5.4 g, 0.036-0.050 mol) and leave refluxing for 4 days (see also descriptions in references) [16, 17]. After cooling to room temperature, admit chilled dry diethyl ether (~20 mL) to the reaction pot and collect the crystals of $[Et_3S]Br$. The isolated product is then washed with small quantities of chilled dry ethanol, then recrystallized from anhydrous ethanol and vacuum dried at 40 °C using the double manifold/Schlenk line. In a dry atmosphere (inert glove box or under positive pressure of dry $Ar_{(g)}$) *very slowly* add quantities of vacuum-sublimed $AlCl_3$ to the full yield of $[Et_3S]Br$, to a finishing molar ratio of 2:1 ($AlCl_3$: $[Et_3S]Br$) with stirring. The resulting liquid is $[Et_3S][Al_2Cl_6Br]$.

2. Reduction of benzophenone to diphenyl methoxide: All the following manipulations are to be performed in an inert glove box or under positive pressure of dry $Ar_{(g)}$. Slowly admit (hydride will react violently with impurity water, bubbling may occur) a two-fold molar equivalent of CaH_2 (based on $AlCl_3$ used in the preparation) to half of the liquid prepared and leave to stir for 12 h. To the other half, add a molar equivalent of benzophenone (based on CaH_2 used) with stirring. Withdraw a $^{13}C\{^1H\}$ NMR sample of the benzophenone solution and record resonances due to the analyte (compare with those reported) [14]. Slowly add the filtered hydridic solution to that containing benzophenone and stir. Again, take a sample for $^{13}C\{^1H\}$ NMR analysis. Note the changes to the observed resonances and compare with those of the literature values [14].

Explanation

The trichlorohydridoaluminate species, [AlCl$_3$H]$^-$, forms after the addition of hydride and then becomes available for the reduction of benzophenone to diphenyl methoxide, as shown in Scheme 4 [13].

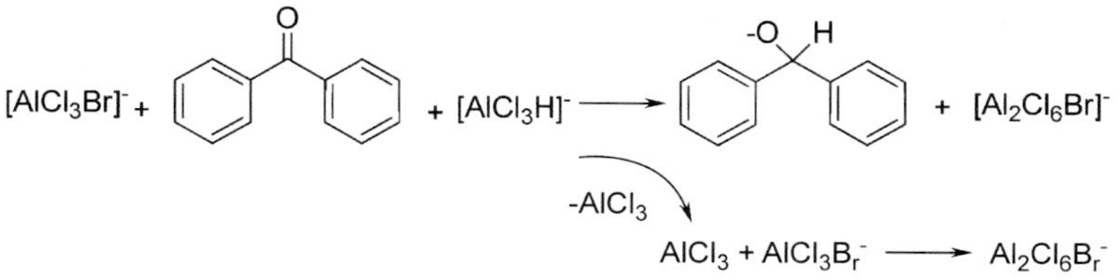

Scheme 4

Aroxide carbanion formation in the ionic liquid demonstrates the unique stabilizing environment that allows for the preparation of nucleophiles that in other solvents would only exist as a transient »intermediate« species. This type of reduction reaction may also provide a synthetic route for the preparation of aldehydes from ketones *via* a »hydro-de(α-oxidoalkyl)-substitution«-like pathway under Brønsted acidic conditions [18] or where the leaving group is able to form a stable carbanion.

References

1 G. Driver, K.E. Johnson, *Green Chem.* **2003**, *5*, 163.

2 G. Driver, M.Sc. thesis, The University of Regina, Regina, Saskatchewan, Canada, **2003**.

3 J.B. Stothers, *Carbon-13 NMR Spectroscopy*, Org. Chem. vol. 24, Academic Press Inc., New York, **1972**.

4 M. Ma, M.Sc. thesis, The University of Regina, Regina, Saskatchewan, Canada, **1993**.

5 F.Y. Fujiwara, J.S. Martin, *J. Chem. Phys.* **1972**, *56*, 4091.

6 R.L. Burwell, *Chem. Rev.* **1954**, *54*, 615.

7 C.A. Smith, J. B. Grutzner, *J. Org. Chem.* **1976**, *41*, 367.

8 S.P. Walvekar, A.B. Halgeri, *Tetrahedron Lett.*, **1971**, *12*, 2663.

9 D. Landini, F. Montanari, F. Rolla, *Synthesis* **1978**, 771.

10 E.C. Friedrich, G. DeLucca, *J. Org. Chem.* **1983**, *48*, 1678.

11 M.E. Jung, M. A. Lyster, *J. Org. Chem.* **1977**, *42*, 3761.

12 M.O.V. Bendetti, E.S. Monteagudo, G.J. Burton, *J. Chem. Research (S)* **1990**, 248.

13 D. Wassell, M.Sc. thesis, The University of Regina, Regina, Saskatchewan, Canada, **1999**.

14 L. Xiao, K.E. Johnson, R.G. Treble, *J. Mol. Cat. A: Chem.* **2004**, *214*, 121.

15 D.F. Wassell, K.E. Johnson, L.M. Mihichuk, *J. Phys. Chem. B* **2007**, *111*, 13578.

16 L. Xiao, K.E. Johnson, *Can. J. Chem.* **2004**, *82*, 491.

17 I.N. Feit, F. Schadt, J. Lubinkowski, W.H. Saunders, *J. Am. Chem. Soc.* **1971**, *93*, 6606.

18 M.B. Smith, J. March, *March's Advanced Organic Chemistry: Reactions, Mechanisms, and Structure*, 5th ed., John Wiley & Sons, Inc., New York, NY, **2001**.

Part III
High Yield and One-Pot Syntheses

21

Efficient Syntheses of Li[Al(OR^F)_4], Ag[Al(OR^F)_4] (R^F = C(CF_3)_3, C(H)(CF_3)_2, C(Me)(CF_3)_2) and [H(OEt_2)_2]^+[Al(OC(CF_3)_3)_4]^-

Ines Raabe, Andreas Reisinger and Ingo Krossing

Until the late 1970s, the term »noncoordinating anion« was used when a small anion like a halide was replaced by a larger complex anion such as BF_4^-, $CF_3SO_3^-$, ClO_4^-, AlX_4^- or MF_6^- (X = F – I, M = P, As, Sb, etc.). But with the advances in structure determination – mainly when single-crystal X-ray diffraction became a routine method – it became obvious that also these complex anions can easily be coordinated if only partnered with a suitable counterion [1], and therefore the term »noncoordinating« became evidently inadequate for these anions. In the early 1990s, the expression »weakly coordinating anion« (WCA) was created, which more accurately describes the interaction between those anions and their counterions, but already includes the potential of such complexes to serve as a precursor of the »noncoordinated« cation, i.e. in catalytic processes. During the last decade, many efforts were undertaken to finally reach the ultimate goal of a truly »noncoordinating anion«. However, noncoordination is physically impossible, but because of the importance of such WCAs both in fundamental [2,3] and applied [4] chemistry, plenty of new, also called »superweak anions« [5], were developed. $[B(C_6F_5)_4]^-$ [6], $[Sb(OTeF_5)_6]^-$ [7], $[CB_{11}Me_6X_6]^-$ [3, 8], or $[Al(OC(CF_3)_3)_4]^-$ [9–11] are some of the most common examples of a new generation of rather large and chemically robust WCAs. Overall, one can exchange a few strongly coordinating interactions for a series of many but very weakly coordinating interactions. Thus, WCAs allow stabilization of strongly acidic gas phase species, highly electrophilic metal and non-metal cations or weakly bound Lewis acid base complexes of metal cations [2–4, 12]. But apart from being very useful in fundamental chemistry, WCAs have a very strong standing in applied chemistry. Thus, WCAs are important for homogeneous catalysis and polymerizations [4, 13, 14], electrochemistry [15], ionic liquids [16], photolithography [17], lithium ion batteries, super capacitors and much more [18]. One important class of WCAs are the poly- and perfluorinated alkoxyaluminates of the type $[Al(OR^F)_4]^-$ [14] in which the univalent negative charge is delocalized over a large number of up to 36 aliphatic C–F bonds. In contrast to the normally easily hydrolyzed alkoxyaluminates, the $[Al(OC(CF_3)_3)_4]^-$ anion is stable in nitric acid (6 mol L^{-1}) [10]. This stability towards hydrolysis was attributed to the steric shielding provided by the bulky $C(CF_3)_3$ ligand, towards the oxygen atoms as well as to the electronic stabilization due to the perfluorination. One of the major advantages of these aluminates is that they are easily accessible on a preparative scale: they can be prepared with little synthetic effort on a 250 g scale with well over 95% yield within two days in common inorganic/organometallic laboratories [11]. The syntheses of important starting materials including the $[Al(OR^F)_4]^-$ WCA are described in this contribution.

Laboratory syntheses of some weakly coordinating anion salts

Preparation of Li[Al(OR^F)$_4$]

Apparatus	Purification of LiAlH$_4$: Extraction frit, 500 mL Schlenk vessel, magnetic stirrer, reflux condenser, oil bath, silicon oil bubbler.
	Synthesis of Li[Al(OR^F)$_4$]: Two necked Schlenk vessel, magnetic stirrer, heating mantle, reflux condenser or, better, gas cooler, cryostat, dropping funnel, silicon oil bubbler. Safety glasses, laboratory coat, protective gloves.
Chemicals	LiAlH$_4$, Et$_2$O, R^{F}OH (with R^F = C(CF$_3$)$_3$, C(H)(CF$_3$)$_2$, C(Me)(CF$_3$)$_2$), purchased from P&M Invest, Russia (http://www.fluorine.ru), petroleum ether (b.p. 60–70 °C; alternatively hexane or heptane).
Attention!	Safety glasses and protective gloves must be worn at all times.
Caution!	All reactions should be carried out in a well-ventilated hood.
	LiAlH$_4$ is flammable and reacts violently with water. Avoid contact with skin and eyes. In case of fire, use sand, CO$_2$ or powder. Never use water. Keep container tightly closed and dry.
	Et$_2$O is extremely flammable, may form explosive peroxides, and is harmful if swallowed. Repeated exposure may cause skin dryness or cracking. Vapors may cause drowsiness and dizziness. Keep container in a well-ventilated place and away from sources of ignition.

Do not empty into drains. No smoking! Take precautionary measures against static discharges.

$(CF_3)_3COH$ is toxic by inhalation, in contact with skin, and if swallowed. Wear suitable protective clothing, gloves and eye/face protection. In case of insufficient ventilation, wear suitable respiratory equipment. In case of accident or if you feel unwell, seek medical advice immediately (show label where possible). In case of accident by inhalation, remove casualty to fresh air and keep at rest.

$(CF_3)_2(H)COH$ is harmful by inhalation and if swallowed, is corrosive and causes burns. In case of contact with eyes, rinse immediately with plenty of water and seek medical advice. Wear suitable protective clothing, gloves and eye/face protection. In case of accident or if you feel unwell, seek medical advice immediately (show label where possible).

$(CF_3)_2(Me)COH$ is harmful by inhalation, in contact with skin and if swallowed. In case of contact with eyes, rinse immediately with plenty of water and seek medical advice. Wear suitable protective clothing, gloves and eye/face protection. In case of accident or if you feel unwell, seek medical advice immediately (show label where possible).

Petroleum ether is extremely flammable, harmful and dangerous for the environment, toxic to aquatic organisms, and may cause long-term adverse effects in the aquatic environment or lung damage if inhaled. Repeated exposure may cause skin dryness or cracking. Vapors may cause drowsiness and dizziness. Keep container in a well-ventilated place and away from sources of ignition. Do not empty into drains. No smoking! Take precautionary measures against static discharges. May cause harm to the unborn child. Possible risk of impaired fertility.

Procedure

Because of the air and moisture sensitivity of most materials, all manipulations were undertaken using standard vacuum (better than 5×10^{-2} mbar in the vacuum line) and Schlenk techniques as well as in a glove box with an argon or nitrogen atmosphere (H_2O and O_2 < 1 ppm). All solvents were dried by conventional drying agents [19] and distilled afterwards. Their residual moisture content was found by Karl Fischer titration to be about 5 ppm. The fluorinated alcohols were distilled and degassed prior to use.

Purification of $LiAlH_4$: Commercially available $LiAlH_4$ is purified by extraction with Et_2O in an extraction frit, which yields a white

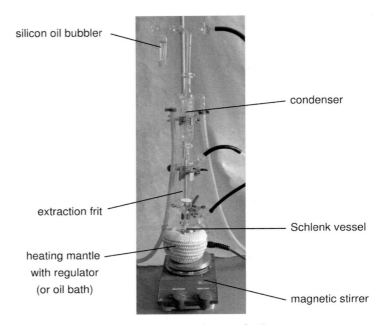

silicon oil bubbler

condenser

extraction frit

Schlenk vessel

heating mantle
with regulator
(or oil bath)

magnetic stirrer

Figure 1   Experimental set-up for the purification of LiAlH$_4$.

powder [19]. All traces of Et$_2$O are removed by weighing the sample constant *in vacuo* (approx. 10^{-3} mbar) with heating to 80 °C (Caution: Do not heat to higher temperatures! Use a water bath, since LiAlH$_4$ explodes at about 120 °C). Usually the LiAlH$_4$ cake has to be ground very finely inside a glove box and is then again exposed to vacuum prior to reaching a constant weight. Depending on the quality of the LiAlH$_4$, up to 30% of the starting material are impurities (Al, Al oxides and Li$_3$AlH$_6$), which remain insoluble on the frit plate.

Synthesis of Li[Al(OR^F)$_4$]: In the glove box, 1 equivalent of purified, *very finely ground* LiAlH$_4$ is weighed into the two-necked Schlenk vessel. After the addition of petroleum ether (approx. 50 mL petrol ether per 1g LiAlH$_4$), the intensive gas cooler with the bubbler as well as the dropping funnel are connected to the Schlenk vessel. Because of the very low boiling points of the alcohols, especially (CF$_3$)$_3$COH (b.p. 45 °C), an intensive reflux condenser or better gas cooler connected to a cryostat and set to a temperature of –25 °C has to be used to avoid evaporation. The fluorinated alcohol (4.1 equivalents) is then added drop by drop to the LiAlH$_4$/petroleum ether suspension. After all the alcohol has been added, the reaction mixture is heated under reflux for about 3 h. The Schlenk vessel is then cooled down to −25 °C to allow complete precipitation of the product. The supernatant solvent is decanted off and the remaining solvent incorporated in the colorless microcrystalline precipitate is then removed *in vacuo* (10^{-3} mbar) until constant weight of the sample (isolated yield: approx. 95–99%) [11].

Characterization data:

Li[Al(OC(H)(CF$_3$)$_2$)$_4$]: m.p. 120–125 °C. ^{1}H NMR (250 MHz, CDCl$_3$/5%THF, 25 °C): δ 4.19 (sept, $^3J_{HF}$ = 6.2 Hz, CH) ppm. ^{7}Li NMR (117 MHz, CDCl$_3$/5%THF, 25 °C): δ −1.1 ppm. ^{13}C NMR (63 MHz, CDCl$_3$/5%THF, 25 °C): δ 70.4 (sept, $^2J_{CF}$ = 32.3 Hz, O-C), 122.7 (q, J_{CF} = 285.0 Hz, CF$_3$) ppm. ^{27}Al NMR (78 MHz, CDCl$_3$/5%THF, 25 °C): δ 59.9 (s, $\omega_{1/2}$ = 230 Hz) ppm. FTRaman (cm^{-1}): $\bar{v}$ (%) 2955 (100), 1390 (15), 1295 (10), 1200 (11), 1129 (4), 1098 (7), 855 (82), 766 (28), 750 (18), 730 (7), 703 (10), 689 (9), 533 (16), 523 (15), 484 (4), 330 (41), 298 (10), 218 (13), 120 (8). IR (Diamond ATR, cm^{-1}): $\bar{v}$ 320 (mw), 379 (mw), 427 (mw), 470 (w), 521 (mw), (537 (w), 563 (w), 577 (w), 689 (ms), 748 (w), 767 (w), 790 (w), 828 (w), 864 (m), 896 (m), 1102 (s), 1136 (vs), 1188 (vs), 1237 (vs), 1293 (s), 1379 (ms).

Li[Al(OC(Me)(CF$_3$)$_2$)$_4$]: m.p. 42–45 °C. ^{1}H NMR (250 MHz, CDCl$_3$, 25 °C): δ 1.57 (s, CH_3) ppm. ^{7}Li NMR (117 MHz, CDCl$_3$, 25 °C): δ −1.0 ppm. ^{13}C NMR (63 MHz, CDCl$_3$, 25 °C): δ 17.2 (s, CH$_3$), 75.8 (sept, $^2J_{CF}$ = 29.6 Hz, O-C), 124.1 (q, J_{CF} = 287.4 Hz, CF$_3$) ppm. ^{27}Al NMR (78 MHz, CDCl$_3$, 25 °C): δ 46.6 (s, $\omega_{1/2}$ = 620 Hz) ppm. FT-Raman (cm^{-1}): $\bar{v}$ (%) 2964 (32), 2862 (100), 774 (15), 543 (33), 332 (5), 247 (4). IR (Diamond ATR, cm^{-1}): $\bar{v}$ 368 (w), 396 (mw), 443 (w), 512 (w), 534 (mw), 572 (w), 633 (w), 702 (ms), 739 (w), 773 (w), 794 (w), 869 (mw), 981 (mw), 995 (mw), 1087 (vs), 1121 (s), 1196 (vs), 1230 (vs), 1311 (s) 1392 (mw), 1460 (mw).

Li[Al(OC(CF$_3$)$_3$)$_4$]: m.p. 145–150 °C. ^{7}Li NMR (117 MHz, CDCl$_3$, 25 °C): δ −0.9 ppm. ^{13}C NMR (63 MHz, CDCl$_3$, 25 °C): δ 120.9 (q, J_{CF} = 292.8 Hz, CF$_3$) ppm. ^{19}F NMR (235 MHz, CDCl$_3$, 25 °C): δ. 76.9 (s, CF_3) ppm. ^{27}Al NMR (78 MHz, CDCl$_3$, 25 °C): δ 33.8 (s, $\omega_{1/2}$ = 130 Hz) ppm. FT-Raman (cm^{-1}): $\bar{v}$ (%) 801 (70), 745 (90), 571 (25), 538 (40), 326 (100), 234 (20). IR (Diamond ATR, cm^{-1}): $\bar{v}$ 289 (w), 326 (w), 369 (w), 464 (m), 539 (mw), 546 (mw), 572 + 582 (m), 726 (s), 740 (ms), 756 + 760 (m), 798 (m), 844 (ms), 863 (ms), 936 (ms), 964 (vs), 976 (vs), 1184 (ms), 1225 (vs), 1243 (s), 1270 (s), 1297 (s), 1353 (ms).

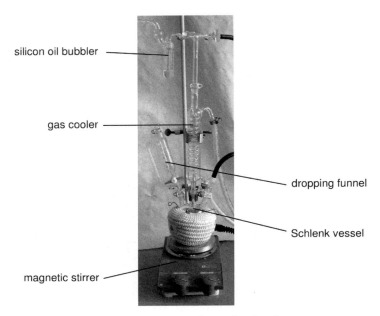

silicon oil bubbler

gas cooler

dropping funnel

Schlenk vessel

magnetic stirrer

Figure 2   Experimental set-up for the synthesis of Li[Al(OR^F)$_4$].

Especially when doing large scale reactions (above 50 g), it might happen that the LiAlH$_4$ does not completely react. To overcome this, very finely ground LiAlH$_4$ is required. If residual Al–H bonds are in the reaction mixture, the white powder reacts noisily with water (Add approximately 100 mg to water. If a clear solution without hissing noise results, the product is okay.) If residual Al–H bonds are present, grind the solid material in the glove box, put it in a Schlenk flask with petroleum ether, add further alcohol and reflux until no further hydrogen evolution is observed.

Waste Disposal   Dispose of Et$_2$O to organic, ether-containing solvent waste (or reuse after distillation).

Residual material on the frit plate (Caution: pyrophoric material!) has to be hydrolyzed slowly. Step-by-step hydrolyzation: small amounts of the residual material are put in a big beaker with ice water. H$_2$ formation! Only to be performed in a well-ventilated fume hood!

R^{F}OH to organic solvent waste (or reuse after distillation).

Petroleum ether to organic solvent waste (or reuse after distillation).

Explanation

Li[Al(OR^F)$_4$] (with R^F = C(CF$_3$)$_3$, C(H)(CF$_3$)$_2$, C(Me)(CF$_3$)$_2$) was formed by reaction of purified LiAlH$_4$ and the appropriate commercially available alcohols [9-11].

$$LiAlH_4 + 4\ R^FOH \xrightarrow{\text{petrol ether}} Li[Al(OR^F)_4] + 4\ H_2$$

Because of the formation of four equivalents of H$_2$, the reaction should only be performed in a well-ventilated fume hood that quickly reduces the hydrogen level and keeps it well below the lower explosion limit.

Preparation of Ag[Al(OR^F)$_4$]

Apparatus

Two-bulbed Schlenk vessel with Young valves and G4 frit plate (Figure 4), magnetic stirrer, Schlenk flask, ultrasonic bath, safety glasses, laboratory coat, protective gloves.

Chemicals

Li[Al(OR^F)$_4$], AgF (best success, if purchased from Apollo Scientific, UK; has to be yellow-orange; brownish-blackish coloration has to be avoided; if in doubt, recrystallize from water and dry at 120 °C in an oven), CH$_2$Cl$_2$.

Attention!

Safety glasses and protective gloves must be worn at all times.

Caution!

All reactions should be carried out in a well-ventilated hood.

AgF is toxic by inhalation, in contact with skin and if swallowed. Corrosive and causes burns. In case of contact with eyes, rinse immediately with plenty of water and seek medical advice. After contact with skin, wash immediately with plenty of water. Wear suitable protective clothing, gloves and eye/face protection. In case of accident or if you feel unwell, seek medical advice immediately (show label where possible).

Li[Al(OR^F)$_4$] hazards and risks unknown.

CH$_2$Cl$_2$ shows limited evidence of a carcinogenic effect. Do not breathe gas/fumes/vapor/spray (appropriate wording to be specified by the manufacturer). Avoid contact with skin and eyes. Wear suitable protective clothing and gloves.

Procedure   Because of air and moisture sensitivity of most materials, all manipulations should be undertaken using standard vacuum (better than 5×10^{-2} mbar in the vacuum line) and Schlenk techniques or in a glove box with an argon or nitrogen atmosphere (H_2O and O_2 < 1 ppm). CH_2Cl_2 was dried by conventional drying agents [19] and distilled afterwards (residual water about 5 ppm by Karl Fischer titration).

In the glove box, the lithium salt and 1.3 equivalents of AgF are weighed into one side of the two-bulbed Schlenk vessel closed by Young valves. The reagents are suspended in CH_2Cl_2 (approx. 50 mL per 10 g $Li[Al(OR^F)_4]$); the vessel is then evacuated until the solvent starts to boil and left under the vapor pressure of the solvent. The mixture is sonicated in a strong ultrasonic bath (e.g. Bandelin Sonorex RM 16 U) overnight. Search for the region of most ultrasonic intensity – the reaction may be slow if the ultrasonic bath is not strong enough or the flask is placed at an area of insufficient ultrasonic intensity. The reaction mixture turns slightly brownish with only little of a dark brown (almost black) precipitate. Prior to filtration, the solution is stored in a −20 to −30 °C freezer for at least 3 h to check whether unreacted $Li[Al(OR^F)_4]$ is still present (formation of a white, microcrystalline precipitate). If no precipitate forms, the reaction is finished and the solution is filtered. The product is then dried *in vacuo*, finely ground in the glove box, placed in a new flask and left directly hooked to a vacuum line (no tubing; directly connect the vessel with the glass joint to the vacuum line) until constant weight is achieved. With this procedure, one gets rid of the last traces of coordinated solvent and obtains solvent-free »naked« $Ag[Al(OR^F)_4]$ (isolated yield: approx. 95%) [10, 11]. *For $R^F = C(CF_3)_3$, it is rather difficult to remove the last traces of CH_2Cl_2.* If the flask is only evacuated through a tubing and *not* directly hooked to a good vacuum line, one usually only gets down to $Ag(CH_2Cl_2)[Al(OR^F)_4]$; if one continues to work with CH_2Cl_2, this is suitable. However, if one wishes to change solvents later, it is recommended to follow the above instructions. Usually it takes about 24 to 72 h (depending on the batch size) for all traces of CH_2Cl_2 to be removed. The success of CH_2Cl_2 removal can be monitored by IR (see Figure 3).

Upon complete removal of CH_2Cl_2, several bands split (443 → 436 and 468; 572 → 566 and 575; 833 → 796, 827 and 862; ... cm^{-1}). Overall the bands get broader of the »naked« $Ag[Al(OR^F)_4]$.

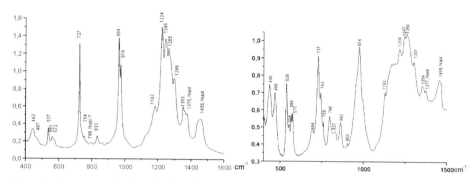

Figure 3 IR spectra of Ag(CH$_2$Cl$_2$)[Al(OR^F)$_4$] (left) and Ag[Al(OR^F)$_4$] (right) (R^F = C(CF$_3$)$_3$).

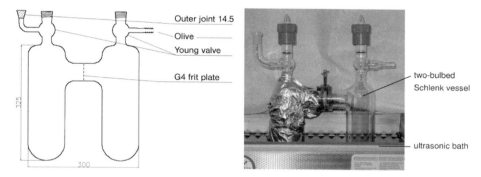

Outer joint 14.5
Olive
Young valve

G4 frit plate

two-bulbed
Schlenk vessel

ultrasonic bath

Figure 4 Drawing of the flask and experimental set-up for the synthesis of Ag[Al(OR^F)$_4$].

Characterization data:

Ag[Al(OC(H)(CF$_3$)$_2$)$_4$]: m.p. 155–160 °C (decomp.). ^{1}H NMR (250 MHz, CDCl$_3$, 25 °C): δ 4.48 (sept, $^3J_{HF}$ = 5.5 Hz, CH) ppm. ^{13}C NMR (63 MHz, CDCl$_3$, 25 °C): δ 75.9 (sept, $^2J_{CF}$ = 32.9 Hz, O–C), 122.1 (q, J_{CF} = 283.1 Hz; CF$_3$) ppm. ^{27}Al NMR (78 MHz, CDCl$_3$, 25 °C): δ 58.0 (s, $\omega_{1/2}$ = 255 Hz) ppm. IR (Diamond ATR, cm^{-1}): $\bar{v}$ 332 (w), 421 (mw), 463 (w), 523 (mw), 537 (w), 563 (w), 578 (w), 669 (w), 688 (m), 727 (w), 760 (mw), 780 (mw), 793 (w), 859 (m), 895 (m), 1091 (vs), 1140 (s), 1187 (vs), 1216 (vs), 1228 (vs), 1259 (vs), 1288 (s), 1377 (ms), 2726 (w), 2951 (mw). Calculated (found) for C$_{12}$H$_4$AgAlF$_{24}$O$_4$: C, 17.95 (18.14); H 0.50 (0.69); Al 3.4 (3.3)%.

[Ag(CH$_2$Cl$_2$)Al(OC(Me)(CF$_3$)$_2$)$_4$]: m.p. 137–145 °C (decomp.). ^{1}H NMR (250 MHz, CDCl$_3$, 25 °C): δ 1.59 (s, 12 H, CH_3), 5.29 (s, 2H, CH_2Cl$_2$) ppm. ^{13}C NMR (63 MHz, CDCl$_3$, 25 °C): δ 17.7 (s, CH$_3$), 75.9 (sept, $^2J_{CF}$ = 29.7 Hz, O-C), 123.8 (q, J_{CF} = 288.2 Hz, CF$_3$) ppm.

^{27}Al NMR (78 MHz, CDCl$_3$, 25 °C): δ 45.2 (s, $\omega_{1/2}$ = 270 Hz) ppm. IR (Diamond ATR, cm^{-1}) of the solvent-free Ag[Al(OC(Me)(CF$_3$)$_2$)$_4$]: $\bar{\nu}$ 371 (w), 392 (w), 439 (mw), 455 (w), 518 (w), 535 (m), 572 (m), 621 (m), 632 (m), 702 (ms), 739 (m), 772 (m), 791 (m), 800 (m), 867 (mw), 998 (mw), 1084 (vs), 1118 (s), 1185 (s), 1210 (vs), 1229 (vs), 1311 (s), 1460 (mw), 2962 (ms), 3012 (mw). Calculated (found) for C$_{17}$H$_{14}$AgAlCl$_2$F$_{24}$O$_4$: Ag, 11.4 (11.9); Al, 2.9 (2.6)%.

[Ag(CH$_2$Cl$_2$)Al(OC(CF$_3$)$_3$)$_4$]: m.p. 97–100 °C (decomp.). ^{1}H NMR (250 MHz, CDCl$_3$, 25 °C): δ 5.34 (s, CH_2Cl$_2$) ppm. ^{13}C NMR (63 MHz, CDCl$_3$, 25 °C): δ 54.0 (s, CH$_2$Cl$_2$), 121.2 (q, J_{CF} = 292.8 Hz, CF$_3$) ppm. ^{27}Al NMR (78 MHz, CDCl$_3$, 25 °C): δ 34.1 (s, $\omega_{1/2}$ = 39 Hz) ppm. IR (Diamond ATR, cm^{-1}): 288 (w), 316 (m), 332 (w), 367 (w), 387 (mw), 443 (m), 468 (mw), 537 (mw), 561 (mw), 572 (mw), 727 (ms), 754 (w), 796 (w), 833 (mw), 964 (s), 974 (ms), 1182 (m), 1224 (vs), 1245 (vs), 1256 (s), 1299 (ms), 1355 (mw). IR (Diamond ATR, cm^{-1}) of the solvent-free Ag[Al(OC(Me)(CF$_3$)$_2$)$_4$]: $\bar{\nu}$ 290 (w), 319 (m), 332 (w), 391 (mw), 435 (m), 468 (m), 538 (ms), 553 (mw), 567 (mw), 575 (mw), 694 (w), 727 (s), 743 (m), 759 (w), 796 (mw), 827 (w), 862 (mw), 974 (vs), 1182 (m), 1218 (vs), 1245 (vs), 1259 (vs), 1301 (s), 1354 (m). Calculated (found) for C$_{17}$H$_2$AgAlCl$_2$F$_{36}$O$_4$: C, 17.60 (16.82); H, 0.17 (0.02)%.

Waste Disposal   AgF solid dispose of to fluorine-containing waste.

CH$_2$Cl$_2$ dispose of to organic solvent containing halide waste (or reuse after distillation).

Explanation

Syntheses of the silver salts are performed by reaction of the lithium salts Li[Al(OR^F)$_4$] obtained by the procedure described above with AgF in CH$_2$Cl$_2$.

$$\text{Li[Al(OR}^F\text{)}_4] + \text{AgF} \xrightarrow{\text{CH}_2\text{Cl}_2} \text{Ag[Al(OR}^F\text{)}_4] + \text{LiF}$$

Preparation of [H(OEt$_2$)$_2$][Al(OC(CF$_3$)$_3$)$_4$]

Apparatus   Schlenk vessel, liquid nitrogen bath, magnetic stirrer, gas balloon, silicon oil bubbler, safety glasses, laboratory coat, protective gloves.

| **Chemicals** | Et_2O, dry HCl gas, $Li[Al(OC(CF_3)_3)_4]$, CH_2Cl_2, toluene. |

Attention!   Safety glasses and protective gloves must be worn at all times.

Caution!   All reactions should be carried out in a well-ventilated hood.

Et_2O: see section above on preparation of $Li[Al(OR^F)_4]$.

HCl is toxic by inhalation and causes severe burns. Keep container in a well-ventilated place. In case of contact with eyes, rinse immediately with plenty of water and seek medical advice. Wear suitable protective clothing, gloves and eye/face protection. In case of accident or if you feel unwell, seek medical advice immediately (show label where possible).

$Li[Al(OR^F)_4]$: see section above on preparation of $Ag[Al(OR^F)_4]$.

CH_2Cl_2: see section above on preparation of $Ag[Al(OR^F)_4]$.

Toluene is highly flammable. Irritating to the skin. Harmful: danger of serious damage to health by prolonged exposure through inhalation. Possible risk of harm to the unborn child. May cause lung damage if inhaled. Vapors may cause drowsiness and dizziness. Wear suitable protective clothing and gloves. If swallowed, do not induce vomiting: seek medical advice immediately and show container or label.

Procedure   Because of air and moisture sensitivity of most materials, all manipulations should be undertaken using standard vacuum (better than 5×10^{-2} mbar in the vacuum line) and Schlenk techniques in a glove box with an argon or nitrogen atmosphere (H_2O and $O_2 < 1$ ppm). CH_2Cl_2 was dried by conventional drying agents [19] and distilled afterwards (residual water about 5 ppm by Karl Fischer titration).

In the glove box, $Li[Al(OC(CF_3)_3)_4]$ is weighed into a Schlenk vessel and then suspended in CH_2Cl_2 (approx. 500 mL per 20 g) and Et_2O (2.5 eq.). The suspension is cooled down with a liquid nitrogen bath and a preweighed amount of gaseous HCl (1.05 eq.) from a gas transfer flask (»gas balloon«, see picture) of known volume is condensed directly onto the cold mixture. The mixture is allowed to reach room temperature and stirred overnight, whereby colorless LiCl is formed. Note that the flask is left under the vapor pressure of the solvent and therefore has to be tested for leak-tightness prior to use. The LiCl precipitate is filtered off (if it is too fine and goes through a G4 frit plate, one should remove all the volatiles and dissolve the residue in CH_2Cl_2 before filtering), leading to a clear, colorless solution of $[H(OEt_2)_2][Al(OC(CF_3)_3)_4]$ in CH_2Cl_2. After removal of all volatiles *in vacuo* from the filtrate, pure $[H(OEt_2)_2][Al(OC(CF_3)_3)_4]$ is isolated as a white powder (isolated yield: approx. 90%). Larger crystals suitable for X-ray crystallography may be obtained by recrystallization of the powder from a mixture of toluene/Et_2O (10:1) [1, 20]. With this procedure one may prepare up to 50 g batches of $[H(OEt_2)_2][Al(OC(CF_3)_3)_4]$.

Waste Disposal Et_2O: see section above on preparation of $Li[Al(OR^F)_4]$.

LiCl dispose of to solid waste

CH_2Cl_2: see section above on preparation of $Ag[Al(OR^F)_4]$.

Toluene dispose of to organic solvent waste (or reuse after distillation)

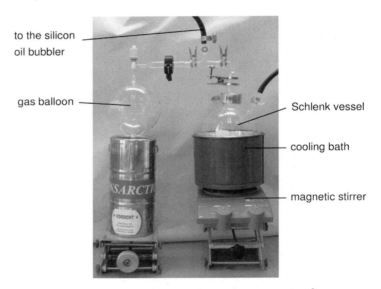

to the silicon oil bubbler

gas balloon

Schlenk vessel

cooling bath

magnetic stirrer

Figure 5   Experimental set-up for the synthesis of $[H(Et_2O)_2][Al(OR^F)_4]$.

Explanation

$Li[Al(OC(CF_3)_3)_4]$ is reacted with gaseous HCl in Et_2O to give the salt of the protonated ether, $[H(OEt_2)_2][Al(OC(CF_3)_3)_4]$.

$$Li[Al(OR^F)_4] + HCl_{(g)} + 2\ Et_2O \longrightarrow [H(OEt_2)_2]^+[Al(OR^F)_4]^- + LiCl$$

References

1 W. Beck, K. Sünkel, *Chem. Rev.* **1988**, *88*, 1405.

2 I. Krossing, I. Raabe, *Angew. Chem.* **2004**, *116*, 2116; *Angew. Chem. Int. Ed.* **2004**, *43*, 2066; S.H. Strauss, *Chem. Rev.* **1993**, *93*, 927.

3 C.A. Reed, *Acc. Chem. Res.* **1998**, *31*, 133.

4 E.Y.-X. Chen, T. J. Marks, *Chem. Rev.* **2000**, *100*, 1391.

5 A.J. Lupinetti, S.H. Strauss, *Chemtracts* **1998**, *11*, 565.

6 A.G. Massey, A.J. Park, *J. Organomet. Chem.* **1964**, *2*, 245.

7 (a) T.S. Cameron, I. Krossing, J. Passmore, *Inorg. Chem.* **2001**, *40*, 4488; (b) D.M. van Seggen, P.K. Hurlburt, O.P. Anderson, S.H. Strauss, *Inorg. Chem.* **1995**, *34*, 3453.

8 D. Stasko, C.A. Reed, *J. Am. Chem. Soc.* **2002**, *124*, 1148.

9 S.M. Ivanova, B.G. Nolan, Y. Kobayashi, S.M. Miller, O.P. Anderson, S.H. Strauss, *Chem.Eur. J.* **2001**, *7*, 503.

10 I. Krossing, *Chem. Eur. J.* **2001**, *7*, 490.

11 I. Krossing, A. Reisinger, *Coord. Chem. Rev.* **2006**, *250*, 2721.

12 (a) E.Y.X. Chen, K.A. Abboud, *Organometallics* **2000**, *19*, 5541; (b) R.E. LaPointe, G.R. Roof, K.A. Abboud, J. Klosin, *J. Am. Chem. Soc.* **2000**, *122*, 9560; (c) V.C. Williams, G.J. Irvine, W.E. Piers, Z. Li, S. Collins, W. Clegg, M.R.J. Elsegood, T.B. Marder, *Organometallics* **2000**, *19*, 1619; (d) J. Zhou, S.J. Lancaster, D.A. Walker, S. Beck, M. Thornton-Pett, M. Bochmann, *J. Am. Chem. Soc.* **2001**, *123*, 223.

13 (a) A. Bell, D. Amoroso, J. Protasiewicz, N. Thirupathi, (Promerus, LLC, USA; Case Western Reserve University). Application: WO, **2005**; (b) C. Böing, G. Franciò, W. Leitner, *Advanced Synthesis and Catalysis* **2005**, *347*, 1537; (c) R.A. Epstein, T.J. Barbarich, (Akzo Nobel NV, Neth.). Application: US, **1999**; (d) P. Hanefeld, V. Böhm, M. Sigl, N. Challand, M. Röper, H.-M. Walter, (BASF Aktiengesellschaft, Germany). Application: WO, **2007**; (e) P. Hanefeld, M. Sigl, V. Böhm, M. Röper, H.-M. Walter, I. Krossing, (BASF Aktiengesellschaft, Germany). Application: WO, **2007**; (f) S. Mihan, (Basell Polyolefine G.m.b.H., Germany). Application: WO, **2005**; (g) L.F. Rhodes, A. Bell, R. Ravikiran, J.C. Fondran, S. Jayaraman, B.L. Goodall, R.A. Mimna, J.-H. Lipian, (Promerus, LLC, USA). Application: US, **2003**; (h) A.J. Rucklidge, D.S. McGuinness, R.P. Tooze, A.M.Z. Slawin, J.D.A. Pelletier, M.J. Hanton, P.B. Webb, *Organometallics* **2007**, *26*, 2782; (i) S.P. Smidt, N. Zimmermann, M. Studer, A. Pfaltz, *Chem. Eur. J.* **2004**, *10*, 4685; (j) S.H. Strauss, B.G. Nolan, T.J. Barbarich, J.J. Rockwell, (Colorado State University Research Foundation, USA). Application: WO, **1999**; (k) S.H. Strauss, B.G. Nolan, B.P. Fauber, (Colorado State University Research Foundation, USA). Application: WO, **2000**; (l) K. Fujiki, J. Ichikawa, H. Kobayashi, A. Sonoda, T. Sonoda, *J. Fluorine Chem.* **2000**, *102*, 293; (m) N.J. Patmore, C. Hague, J.H. Cotgreave, M.F. Mahon, C.G. Frost, A.S. Weller, *Chem. Eur. J.* **2002**, *8*, 2088; (n) S.D. Ittel, L.K. Johnson, M. Brookhart, *Chem. Rev.* **2000**, *100*, 1169; (o) S. Mecking, *Coord. Chem. Rev.* **2000**, *203*, 325; (p) H.H. Brintzinger, D. Fischer, R. Mülhaupt, B. Rieger, R.M. Waymouth, *Angew. Chem.* **1995**, *107*, 1255; *Angew. Chem. Int. Ed.* **1995**, *34*, 1143; (q) W.E. Piers, *Chem. Eur. J.* **1998**, *4*, 13; (r) C. Zuccaccia, N.G. Stahl, A. Macchioni, M.-C. Chen, J.A. Roberts, T.J. Marks, *J. Am. Chem. Soc.* **2004**, *126*, 1448; (s) M.-C. Chen, J.A.S. Roberts, T.J. Marks, *J. Am. Chem. Soc.* **2004**, *126*, 4605; (t) M.-C. Chen, J.A.S. Roberts, T.J. Marks, *Organometallics* **2004**, *23*, 932.

14 T.J. Barbarich, S.M. Miller, O.P. Anderson, S.H. Strauss, *J. Mol. Catal. A* **1998**, *128*, 289.

15 (a) F. Barriere, N. Camire, W.E. Geiger, U.T. Müller-Westerhoff, R. Sanders, *J. Am. Chem. Soc.* **2002**, *124*, 7262; (b) F. Barriere, W.E. Geiger, *J. Am. Chem. Soc.* **2006**, *128*, 3980; (c) N. Camire, U.T. Müller-Westerhoff, W.E. Geiger, *J. Organomet. Chem.* **2001**, *637-639*, 823; (d) N. Camire, A. Nafady, W.E. Geiger, *J. Am. Chem. Soc.* **2002**, *124*, 7260; P.G. Gassman, P.A. Deck, *Organometallics* **1994**, *13*, 1934; (e) P.G. Gassman, J.R. Sowa, Jr., M.G. Hill, K.R. Mann, *Organometallics* **1995**, *14*, 4879; (f) M.G. Hill, W.M. Lamanna, K.R. Mann, *Inorg. Chem.* **1991**, *30*, 4687; (g) R.J. LeSuer, W.E. Geiger, *Angew. Chem.* **2000**, *112*, 254; *Angew. Chem. Int. Ed.* **2000**, *39*, 248; (h) L. Pospísil, B.T. King, J. Michl, *Electrochimica Acta* **1998**, *44*, 103.

16 (a) P. Wasserscheid, W. Keim, *Angew. Chem.* **2000**, *112*, 3926; *Angew. Chem. Int. Ed.* **2000**, *39*, 3772; (b) T. Welton, *Chem. Rev.* **1999**, *99*, 2071; (c) A. Bösmann, G. Francio, E. Janssen, M. Solinas, W. Leitner, P. Wasserscheid, *Angew. Chem.* **2001**, *113*, 2769; *Angew. Chem. Int. Ed.* **2001**, *40*, 2697.

17 (a) F. Castellanos, J.P. Fouassier, C. Priou, J. Cavezzan, *J. Appl. Polymer Sci.* **1996**, *60*, 705; (b) K. Ren, J.H. Malpert, H. Li, H. Gu, D.C. Neckers, *Macromolecules* **2002**, *35*, 1632.

18 (a) N. Ignat'ev, P. Sartori, *J. Fluorine Chem.* **2000**, *101*, 203; (b) F. Kita, H. Sakata, A. Kawakami, H. Kamizori, T. Sonoda, H. Nagashima, N.V. Pavlenko, Y.L. Yagupolskii, *J. Power Sources* **2001**, *97-98*, 581; (c) F. Kita, H. Sakata, S. Sinomoto, A. Kawakami, H. Kamizori, T. Sonoda, H. Nagashima, J. Nie, N.V. Pavlenko, Y.L. Yagupolskii, *J. Power Sources* **2000**, *90*, 27; (d) L.M. Yagupolskii, Y.L. Yagupolskii, *J. Fluorine Chem.* **1995**, *72*, 225.

19 W.L.F. Armarego, C.L.L. Chai, *Purification of Laboratory Chemicals*, 5th ed., Butterworth-Heinemann, Amsterdam, **2003**.

20 I. Krossing, A. Reisinger, *Eur. J. Inorg. Chem.* **2005**, 1979.

22

Quantitative Synthesis of a Neutral Hexacoordinate Phosphorus Compound

Dietmar K. Kennepohl

Neutral compounds can be prepared in which the phosphorus center is six-co-ordinate. This fact is not well known even within the general chemistry com-munity. Anions of phosphorus, such as $[PF_6]^-$, are much more common and fa-miliar examples of the hexacoordinate phosphorus center. Nevertheless, neutral phosphorus compounds with a highly coordinatively saturated phosphorus cen-ter can be accessed in a number of different ways, and a substantial body of lit-erature has appeared on these systems over the years in widely scattered sources [1].

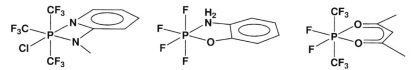

Scheme 1 Some neutral hexacoordinate phosphorus compound examples showing four-, five- and six-membered rings.

These neutral hexacoordinate phosphorus compounds are of interest to re-searchers as model compounds. There is a realization that the highly coordinat-ed phosphorus centers may play an important role in reaction processes involv-ing organophosphorus chemistry [2–4]. In particular, the recognition that nu-cleophilic attack on pentacoordinate and pentavalent phosphorus could well in-volve a six-coordinate intermediate which would be similar to the stable species described here. The highly coordinated phosphorus center is formally »hyper-valent« (i.e. beyond the octet) and so has also been the subject of interest and debate with regard to the question of d-orbital involvement in the bonding.

It is clear that a six-coordinate environment of phosphorus is reasonably ac-cessible in neutral compounds and that a large variety of stable molecular species may be constructed given an appropriate choice of substituents on the central atom. Highly electronegative substituents increase the Lewis acidity of the phosphorus center and so promote intramolecular acid-base adduct formation with a concomitant increase in the central atom coordination. As would be expected, most of these neutral hexacoordinate compounds are derived from pentavalent phosphorus, with only a very few examples of systems based on trivalent phosphorus.

One interesting subclass of these compounds consists of those that have been formed by an apparent insertion of a molecule into one of the bonds of the pentacoordinate precursor. For example, the aminophosphorane $Me(CF_3)_3PNMe_2$ can undergo apparent insertion reactions with CO_2, COS, or CS_2 to form the respective six-coordinate chelated carbamato, monothiocarbamato, and dithiocarbamato complexes $Me(CF_3)_3P[E_2CNMe_2]$, where $CE_2 = CO_2$, COS or CS_2 [5, 6]. In similar fashion, the synthesis of compounds of the type $Cl_{4-n}(CF_3)_nP[(NR)_2CCl]$, where n = 1–3 and R = cyclohexyl or isopropyl are formed by means of an apparent carbodiimide insertion into a P–Cl bond of a pentacoordinate phosphorane [7, 8].

In the experiment described below, we carry out this insertion reaction using phosphorus pentachloride as the starting material. This is not only commercially available, but also less demanding to work with than the analogous fluoro- and fluoromethylphosphoranes. The general form of the reaction is shown in Scheme 2, where R = cyclohexyl or isopropyl.

Scheme 2

Although the reactants and products are not overly reactive with oxygen, they are moisture sensitive. The neutral hexacoordinate phosphorus product will readily hydrolyze to the phosphoryl compound when exposed to air (Scheme 3). It forms a thick yellow liquid with a ^{31}P NMR chemical shift of δ –10.5 ppm ($Cl_3P{=}O$ typically resonates at –2 ppm) [9].

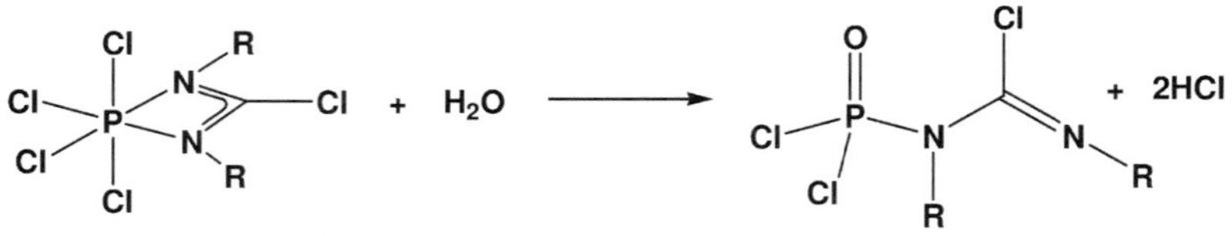

Scheme 3

Preparation of tetrachloro(*N*,*N*'-diisopropylchloroamidino)phosphorus (V)

Apparatus Three-necked round-bottomed flask (100 mL), reflux condenser, syringe (1 mL) with stainless steel needle, spatula, magnetic stir bar, magnetic stirrer, heating mantle, N_2 inlet, Schlenk line, safety glasses, laboratory coat, protective gloves.

Chemicals PCl_5, diisopropylcarbodiimide (DPC), and $CHCl_3$.

Attention! Safety glasses, suitable protective clothing and protective gloves must be worn at all times.

Caution! All reactions should be carried out in a well-ventilated hood.

Diisopropylcarbodiimide (DPC): flammable, very toxic by inhalation, irritating to skin, and can cause serious eye damage. In case of contact with eyes, rinse immediately with plenty of water and seek medical advice. In case of accident or if you feel unwell, seek medical advice immediately (show label where possible). Keep container tightly closed and dry.

Phosphorus pentachloride (PCl_5): reacts violently with water, very toxic by inhalation or if swallowed, and causes burns. In case of contact with eyes, rinse immediately with plenty of water and seek medical advice. In case of accident or if you feel unwell, seek medical advice immediately (show label where possible). Keep container tightly closed and dry.

Trichloromethane ($CHCl_3$): harmful if swallowed and irritating to skin. Danger of serious damage to health by prolonged exposure through inhalation and if swallowed.

Procedure Both reactants and product are hydrolytically sensitive, so all manipulations should be done using Schlenk techniques under an inert atmosphere and in dry glassware. The trichloromethane should be dried by conventional drying agents and distilled afterwards [10].

Charge the 100 mL flask with freshly distilled $CHCl_3$ (50 mL) under a dry inert atmosphere in a round-bottomed flask. Add PCl_5 (ca. 0.5-1.2 g) and dissolve in the $CHCl_3$. To this stirred solution slowly add a stoichiometric amount of DPC by syringe. Gently reflux the mixture under N_2 for 3 h and then cool to room temperature. The solvent is removed in vacuum to leave a white solid.

Characterization data: m.p. 76–78 °C. ^{1}H NMR (400 MHz, CDCl$_3$, 25 °C): δ 1.42 (d, $^3J_{HH}$ = 7 Hz, CH_3CHCH_3), 4.50 (d of sept, $^3J_{HP}$ = 34.8 Hz, CH$_3$CHCH$_3$) ppm. ^{31}P NMR (162 MHz, CDCl$_3$, 25 °C): δ −205.2 (tr, $^3J_{PH}$ = 34.8 Hz) ppm. IR (cm^{-1}): $\bar{v}$ 2990 (s), 2930, 2880, 1668, 1528(vs), 1450, 1387, 1366(s), 1343, 1289, 1212(vs), 1169, 1131, 1122(s), 931, 699, 654(s), 585, 533, 517(vs), 466(vs), 438(vs). MS (70 eV EI): *m/z* 299 (M-Cl, 23.2%), 173 (PCl$_4$, 64.5%), 126 (CN$_2$C$_6$H$_{14}$, 11.1%), 104 (ClCN$_2$C$_2$H$_4$, 43.9%), 83 (CN$_2$C$_3$H$_7$, 12.5%), 69 (CNC$_3$H$_7$, 100%), 43 (C$_3$H$_7$, 90.3%).

Waste Disposal   Dispose of CHCl$_3$ in organic halide solvent waste (or reuse after distillation).

Explanation

The reaction of PCl$_5$ with DPC is quantitative. This, coupled with the fact that the insertion of the carbodiimide is an addition to the phosphorus pentachloride with formally no by-products, makes this reaction a good illustration of the maximization of atom economy. That is, the final product is obtained in high yield and contains the maximum proportion of starting materials.

References

1 C.Y. Wong, D.K. Kennepohl, R.G. Cavell, *Chem. Rev.* **1996**, *96*, 1917.

2 R.R. Holmes, *Chem. Rev.* **1996**, *96*, 927.

3 K.A. Johnson, S.J. Benkovic, In *The Enzymes, 3rd ed.*; D.S. Sigman, P.D. Boyer, Eds.; Academic Press: New York, **1990**, *XIX*, 177.

4 J. Michalski, A. Skowronska, R. Bodalski, Mechanisms of Reactions of Phosphorus Compounds. In *Phosphorus-31 NMR Spectroscopy in Stereochemical Analysis*; J.G. Verkade, L.D. Quin, Eds.; VCH: Deerfield Beach, FL, **1987**, Chapter 8, 255.

5 K.I. The, L. Vande Griend, W.A. Whitla, R.G. Cavell, *J. Am. Chem. Soc.* **1977**, *99*, 7379.

6 R.G. Cavell, K.I. The, L. Vande Griend, *Inorg. Chem.* **1981**, *20*, 3813.

7 D.K. Kennepohl, B.D. Santarsiero, R.G. Cavell, *Inorg. Chem.* **1990**, *29*, 5081.

8 D.K. Kennepohl, R.G. Cavell, *Phosphorus Sulfur Silicon* **1990**, *49/50*, 359.

9 D.E.C. Corbridge, *Phosphorus, 2nd ed.*; Elsevier Scientific Publishing Co.; Amsterdam, **1980**.

10 W.L.F. Armarego, C.L.L. Chai, *Purification of Laboratory Chemicals, 5th ed.*, Butterworth-Heinemann, Amsterdam, **2003**.

23

Encapsulated Silicon in Hexacoordination – Synthesis and Synthetic Transformations

Adinarayana Doddi and M.N. Sudheendra Rao

Silicon is more abundant than its congener, carbon. The chemical behavior of silicon is different from that of carbon because of its greater atomic size. It has played an active role in the development of inorganic, organic and polymer chemistry, and has proved beneficial to the fields of catalysis, biology and materials science [1–5].

One intriguing type of behavior of silicon has been in readily affording higher-coordinate derivatives in both ionic and neutral forms, stabilized especially by using chelating ligands. Such silicon compounds are known to exhibit unusual and interesting features of structure and reactivity.

In recent years, we have developed an easy synthetic route to six-coordinate silicates and have demonstrated their utility in synthetic transformations involving metal and nonmetal substrates (Scheme 1) [6–8].

Syntheses representing all of the above are described below, offering convenient introductory practical laboratory experience of this chemistry.

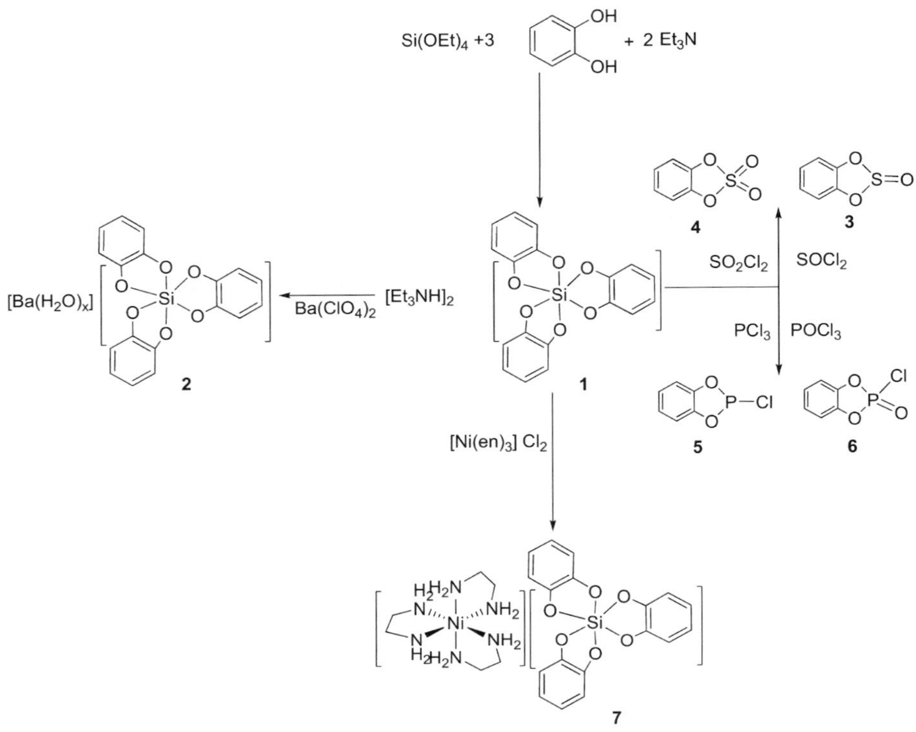

Scheme 1 Ammonium tris(catecholato)silicate and its synthetic transformations.

Some laboratory experiments with encapsulated hexacoordinate silicon compounds

Preparation of bis(triethylammonium)tris(catecholato)silicate (TEASICAT), [(C$_2$H$_5$)$_3$NH]$_2$[Si(O$_2$C$_6$H$_4$)$_3$] (1)

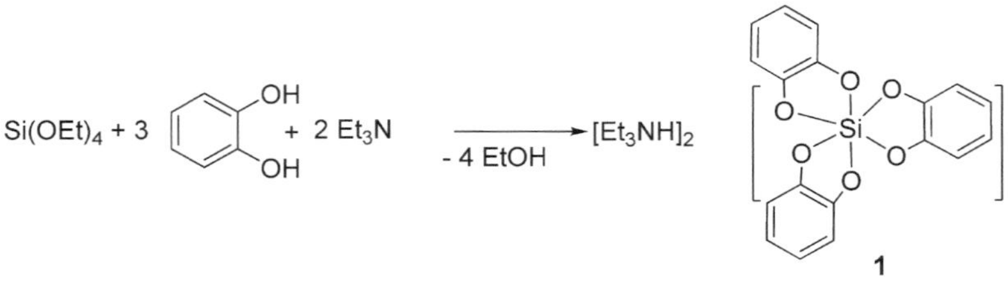

Scheme 2

Apparatus	Flame-dried sidearm round-bottomed flasks (100 mL), syringes (5 mL) with stainless steel needles, rubber septum, magnetic stir bar, magnetic stirrer, sample vials, filtering frit, nickel spatula, a stand and a clamp, safety glasses, laboratory coat, protective gloves.
Chemicals	Tetraethoxysilane (TEOS), acetonitrile (distilled over P_4O_{10} followed by distillation over CaH_2), catechol (purified by sublimation before use), triethylamine (refluxed over CaH_2 and distilled), diethyl ether (dried over $CaCl_2$ followed by distillation over sodium/benzophenone).
Attention!	Safety glasses, laboratory coat and protective gloves must be worn at all times.
Caution!	Triethylamine (flammable liquid, poisonous) should be handled with care in a fume hood. Avoid direct contact with catechol (skin irritant).
Procedure	A flame-dried two-necked round-bottomed flask equipped with a magnetic stir bar, rubber septum and argon or nitrogen inlet (Figure 1) is charged with anhydrous acetonitrile (30 mL), catechol (2.69 g, 26.90 mmol) and triethylamine (2.4 mL, 1.81 g, 18.0 mmol) under an inert atmosphere. To this stirred solution, tetraethoxysilane (2.0 mL, 1.87 g, 9.0 mmol) in a syringe under nitrogen purging is added drop by drop at room temperature as a neat liquid over a period of 15 min. Formation of a white precipitate is observed in about

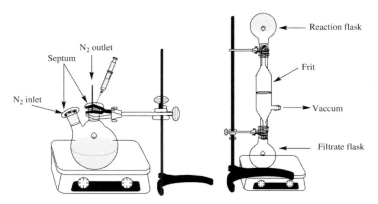

Figure 1 Experimental and filtration set up.

30 min. The reaction mixture is stirred for 2 h and filtered to obtain the title compound as a precipitate, which is washed with ether (3 × 5 mL) and dried (Yield 4.48 g, 90%).

Characterization data: m.p. > 250 °C. ^{1}H NMR (CD$_3$CN): δ 1.15 (t, 18H), 3.10 (q, 12H), 6.35–6.70 (m, 12H) ppm. ^{13}C NMR (CD$_3$CN): δ 12.0, 41.5, 116.3, 120.1, 145.7 ppm. ^{29}Si NMR (CD$_3$CN): δ −139.3 ppm. IR (cm^{-1}): $\bar{v}$ 3021(s), 1594(vs), 1492(vs), 1358(vs), 1249(vs), 1096(vs), 1015(s), 885(vs), 870(s), 815(vs), 738(vs), 687(vs), 590(s), 548(m), 522(vs), 506(m), 443(s).

Waste Disposal   Distill the filtrate to recover acetonitrile, and store in a properly labeled container.

Residues left behind (e.g. unreacted catechol) are disposed of in an alcoholic KOH bath.

Drying agents used are slowly hydrolyzed (since this process is exothermic in nature) and disposed of as basic or acidic wastes.

Explanation

The method is a practical and convenient one-step synthesis of TEASICAT, which is isolable in good overall yield (90%). The compound is air-stable powdery solid, preferably stored in sealed vials. It can be crystallized from acetonitrile at 0 °C. The synthesis can also be performed with many other amines.

Preparation of barium tris(catecholato)silicate (BaSICAT), [Ba(H$_2$O)$_x$][Si(O$_2$C$_6$H$_4$)$_3$] (2)

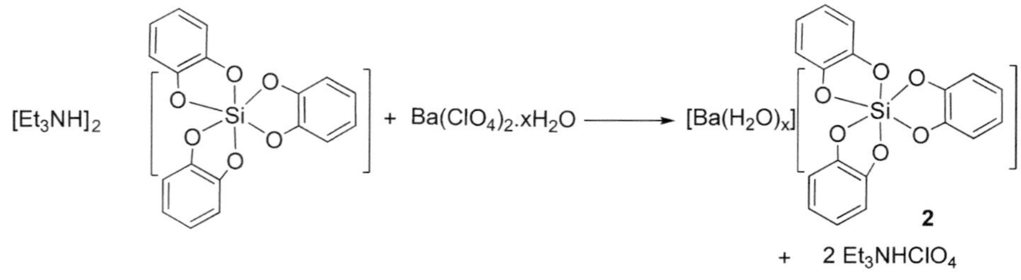

Scheme 3

Apparatus   A flame-dried side-arm round-bottomed flask (100 mL), Syringe (5 mL) with stainless steel needle, septum, magnetic stir bar, magnetic stirrer, sample vials, filtering frit, nickel spatula, porcelain crucible, safety glasses, laboratory coat, protective gloves.

Chemicals   TEASICAT (**1**), acetonitrile (distilled over P_4O_{10} followed by distillation over CaH_2), $Ba(ClO_4)_2$.

Attention!   Safety glasses, laboratory coat, and protective gloves must be worn at all times.

Caution!   Barium perchlorate is explosive and a strong oxidizing agent. Inhalation or contact with eyes or skin causes irritation. Prolonged exposure to fire or heat may result in an explosion. This material must be handled in a fume hood only. Keep material out of water sources and sewers.

Procedure   To a flame-dried two-necked round-bottomed flask containing **1** (1.00 g, 1.79 mmol) in acetonitrile (20 mL, clear solution), a solution of barium perchlorate (0.79 g, 1.78 mmol) in acetonitrile (10 mL) is added by a syringe while stirring at room temperature over a period of 15 min. After half the addition, turbidity appears and by complete addition a good amount of white precipitate is seen. The reaction mixture is stirred for 2 h and filtered through a G-3 frit to obtain the title compound, $[Ba(H_2O)_x][Si(O_2C_6H_4)_3]$ as a white precipitate which is washed with small amounts of acetonitrile and dried (yield 1.0 g, near quantitative).

Characterization data: ^{29}Si NMR (D_2O): δ −144.0 ppm. IR (cm^{-1}): $\bar{v}$ 3589(br), 1594(s), 1489(vs), 1457(m), 1347(s), 1250(vs), 1100(s), 1017(vs), 912(m), 885(s), 869(s), 815(vs), 748(vs), 692(vs), 596(s), 526(vs).

Waste Disposal   Acetonitrile is recovered by distillation and stored in a labeled container.

Residues left behind (e.g. unreacted catechol) are disposed of in an alcoholic KOH bath.

Drying agents used are slowly hydrolyzed (since this process is exothermic in nature) and disposed of as basic or acidic wastes.

Store perchlorate wastes in a separate labeled container.

Explanation

The reaction is very facile at room temperature and the yield is nearly quantitative. The product is stable for a long time in sealed vials and is soluble in methanol. Its IR spectrum clearly reveals the presence of water molecules. IR frequencies seen in the fingerprint region are exclusive to tris(catecholato)silicate ion. This synthetic procedure is also suitable for other group 2 metal ions.

Reactions of bis(triethylammonium)tris(catecholato)silicate with sulfur and phosphorus reagents to produce compounds 3–6

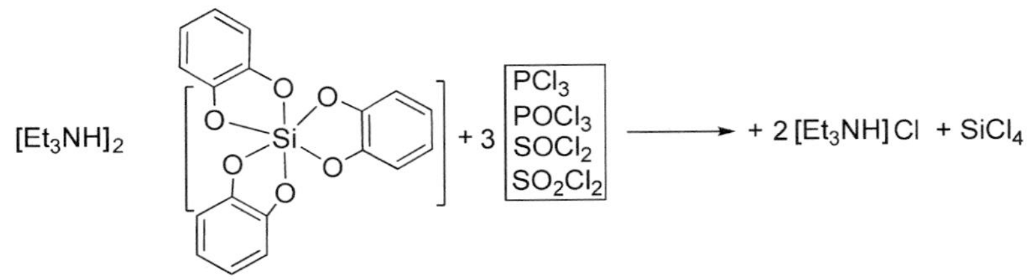

Scheme 4

Apparatus	Magnetic stir bar, magnetic stirrer, sample vials, filtering frit, nickel spatula, side-arm round-bottomed flask (100 mL), syringe (5 mL) with needle, septum, a stand and two clamps, safety glasses, laboratory coat, protective gloves.
Chemicals	Acetonitrile (distilled over P_4O_{10} followed by distillation over CaH_2), hexane (dried over $CaCl_2$ followed by distillation over sodium metal), toluene (dried over $CaCl_2$ followed by distillation over sodium metal), TEASICAT (1), freshly distilled samples of SO_2Cl_2, $SOCl_2$, PCl_3 and $POCl_3$.
Attention!	Handle SO_2Cl_2, $SOCl_2$, PCl_3, $POCl_3$ in a well-ventilated fume hood under inert atmosphere since they are air and moisture sensitive. Safety glasses, laboratory coat, and protective gloves must be worn at all times.
Caution!	SO_2Cl_2: toxic and corrosive, $SOCl_2$: corrosive and lachrymatory, PCl_3: toxic and corrosive, $POCl_3$: toxic and corrosive, hexane: flammable

liquid to be kept away from sources of ignition (e.g. flame, electrical sparks).

Procedure   A flame-dried two-necked round-bottomed flask equipped with a magnetic stir bar, rubber septum and argon or nitrogen inlet is charged with **1** (2.00 g, 3.60 mmol) and acetonitrile (30 mL) and kept stirred. To this clear solution, 1 mL of thionyl chloride (1.30 g, 10.80 mmol) is added drop by drop (by syringe) under N_2 atmosphere over a period of 15 min. After 2 h, the reaction mixture is filtered to isolate the insoluble silica (0.20 g). The solvent is removed in a vacuum to obtain a sticky oily mass, which is extracted with small portions of toluene and hexane (1:1; 4 × 5 mL). Removal of solvent from the extract gives catechol sulfite (**3**) as a yellow liquid (1.51 g). The procedure is repeated with SO_2Cl_2, PCl_3 and $POCl_3$ to get compounds **4**, **5** and **6** respectively.

Characterization data:

(**3**) Yield; 90%. ^{1}H NMR (CDCl$_3$): δ 6.90-7.20(m) ppm. ^{13}C NMR (CDCl$_3$): δ 113.0, 123.2, 147.2 ppm. IR (cm^{-1}): $\bar{v}$ 1230(s).

(**4**) Yield; 90%. m.p. 35–36 °C. ^{1}H NMR (CDCl$_3$): δ 6.90–7.31(m). ^{13}C NMR (CDCl$_3$): δ 113.0, 126.2, 147.3 ppm. IR (cm^{-1}): $\bar{v}$ 1380(s), 1128(br).

(**5**) Yield; 95%. ^{1}H NMR (CDCl$_3$): δ 6.82–7.01(m). ^{13}C NMR (CDCl$_3$): δ 112.0, 121.9, and 146.8. ^{31}P NMR (CDCl$_3$): δ 173.3 ppm. IR (cm^{-1}): $\bar{v}$ 1594(s), 1478(vs), 1408(w), 1347(m), 1328(s), 1258(vs), 1219(vs), 1174(s), 1098(s), 1005(vs), 912(vs), 835(vs), 794(m), 749(vs), 717(s), 618(s).

(**6**) Yield; 85%. ^{1}H NMR (CDCl$_3$): δ 6.92–7.15(m) ppm. ^{13}C NMR (CDCl$_3$): δ 112.1, 123.3 and 145.7 ppm. ^{31}P NMR (CDCl$_3$): δ 18.2(s) ppm. IR (cm^{-1}): $\bar{v}$ 1220(br).

Waste Disposal   Residues left behind are disposed in alcoholic KOH bath.

Solvents recovered are stored in labeled containers.

Drying agents used are slowly hydrolyzed (since this process is exothermic in nature) and disposed of as basic or acidic wastes.

Metallic sodium used for drying solvents is collected and recycled. The residual sodium wastes are carefully hydrolyzed in cold ethanol or isopropanol.

Explanation

The following are the merits of the above reaction: (i) faster reaction, (ii) easy workup procedure and (iii) high yields of products.

Preparation of tris(ethylenediamine)nickel(II)tris(catecholato)silicate [Ni(en)$_3$][Si(O$_2$C$_6$H$_4$)$_3$] (7)

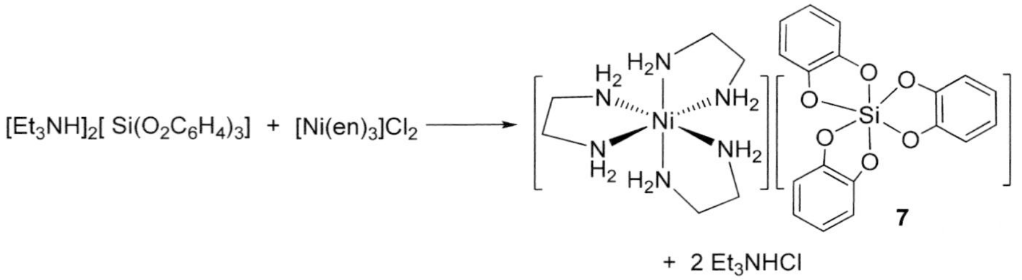

$$[Et_3NH]_2[\,Si(O_2C_6H_4)_3] + [Ni(en)_3]Cl_2 \longrightarrow$$

7

$$+ \ 2 \ Et_3NHCl$$

Scheme 5

Apparatus	Two-necked round-bottomed flasks (100 mL), syringe (5 mL) with needle, septum, magnetic stir bar, magnetic stirrer, sample vials, filtering frit, nickel spatula, safety glasses, laboratory coat, protective gloves.
Chemicals	Acetonitrile (distilled over P$_4$O$_{10}$ followed by distillation over CaH$_2$), diethyl ether (dried over CaCl$_2$ followed by distillation over sodium/benzophenone), TEASICAT (**1**), methanol (dry and distilled).
Attention!	Safety glasses, laboratory coat, and protective gloves must be worn at all times.
Procedure	A flame-dried two-necked round-bottomed flask equipped with a magnetic stir bar, rubber septum and argon or nitrogen inlet is charged with **1** (1.00 g, 1.80 mmol) in acetonitrile (20 mL). To this stirred solution, a methanol solution (10 mL) of [Ni(en)$_3$]Cl$_2$ (0.56 g, 1.80 mmol) is added at room temperature over a period of 15 min. Immediate precipitate formation is observed. After 1 h, the precipitate formed is filtered and washed with methanol (3 × 5 mL) and acetone (4 × 5 mL), and dried *in vacuo* to obtain the title compound (1.04 g, quantitative).

Characterization data: m.p. > 250 °C. Visible (λ_{max} nm): 525. IR (cm^{-1}): $\bar{\nu}$ 3340(s), 3320(s), 3180(w), 1600(m), 1580(m), 1490(vs), 1460(s), 1380(m), 1320(m), 1250(vs), 1140(w), 1100(m), 1020(s), 950(w), 900(w), 880(m), 860(w), 820(s), 740(s), 690(s).

Waste Disposal   Drying agents (CaCl$_2$, CaH$_2$) are slowly hydrolyzed and disposed of as basic or acidic wastes.

Metallic sodium used for drying solvents is collected and recycled. The residual sodium wastes are carefully hydrolyzed in cold ethanol or isopropanol.

Solvents of reaction are recovered and stored in labeled containers.

Explanation

Ion exchange of TEASICAT with [Ni(en)$_3$]Cl$_2$ is almost instantaneous and quantitative. The exchanged product, [Ni(en)$_3$][Si(O$_2$C$_6$H$_4$)$_3$], is highly stable and insoluble in most solvents. The route is applicable to other transition metals.

References

1 D.E. Soli, S.A. Manoso, M. Patterson, P. Deshong, D.A. Favor, R. Hirschmann, B.A. Smith, *J. Org. Chem.* **1999**, *64*, 6526.
2 K. Shu, N. Koichi, *J. Org. Chem.* **1994**, *59*, 6620.
3 B. Delord, M.C. Guillorit, J. Lafay, M.L. Andreola, D. Tharaud, L. Tarrago-Litvak, H.J.A. Fleury, G. Déléris, *Eur. J. Med. Chem.* **1996**, *31*, 111.
4 C.Y. Anderson, K. Freye, K.A. Tubesing, Y.-S. Li, M.E. Kenney, H. Mukhtar, C.A. Elmets, *Photochem. Photobiol.* **1998**, *67*, 332.
5 J. He, H.E. Larkin, Y.S. Li, B.D. Rihter, S.I. A. Zaidi, M.A. J. Rodgers, H. Mukhtar, M.E. Kenney, N.L. Oleinick, *Photochem. Photobiol.* **1997**, *65*, 581.
6 P. Bindu, J.V. Kingston, M.N.S. Rao, *Polyhedron* **2004**, *23*, 679.
7 J.V. Kingston, M.N.S. Rao, *Tetrahedron Lett.* **1997**, *38*, 4841.
8 J.V. Kingston, B. Vargheese, M.N.S. Rao, *Main Group Chemistry* **2000**, *3*, 79.

24

Domino Reactions for the Efficient Synthesis of Natural Products

Lutz F. Tietze, Dirk A. Spiegl and C. Christian Brazel

Over the last 50 years synthetic organic chemistry has developed in a fascinating way. While in the early days only simple molecules could be prepared, nowadays the total synthesis of highly complex natural products has been achieved [1]. The prerequisite for these achievements was the development of highly chemo-, regio- and stereoselective transformations to make an efficient synthesis of complex molecules over a large number of steps possible.

However, a change of paradigm has become apparent over the last decade. Actually, the question is no longer which molecules can be prepared, since limitations according to synthetic methodology have continually decreased, but how to perform the synthesis in an efficient way. Modern synthetic chemistry has to take into account the needs of our environment and sustainability, which especially includes the preservation of resources. Moreover, efficiency in organic synthesis can be defined as the increase of molecular complexity per transformation.

An approach to meet these requirements is the development of domino reactions [2]. A domino reaction according to the definition by Tietze is a process in which two or more bond-forming transformations are carried out under identical reaction conditions, such that the latter transformation takes place at the functionality obtained in the former bond-forming reaction. Thus, a domino reaction is a time-resolved process comparable to domino tiles arranged in a special order, where the first tile, after being tipped over, tips over the next, and the next ... such that they all fall down one after the other.

The ecological as well as economical benefits of forming several bonds in one process are evident, since the amounts of reagents, solvents and waste products can be reduced compared to the classical approach in which bonds are formed individually with workup and purification procedures after each step.

The concept of domino reactions is not a completely new invention, since Nature has already developed several domino reactions for the synthesis of natural

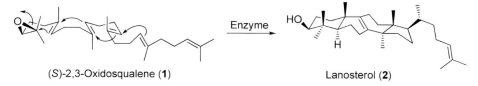

(S)-2,3-Oxidosqualene (**1**)　　　　　　　　　　Lanosterol (**2**)

Scheme 1　Biosynthesis of lanosterol (**2**).

products. For example, in the biosynthesis of steroids the enzymatic cyclization of (S)-2,3-oxidosqualene (**1**) is triggered by opening of the epoxide and followed by the formation of four new bonds and the selective introduction of six stereogenic centers to give lanosterol (**2**) [3].

Domino reactions are classified, again according to Tietze, in terms of the mechanism of the bond-forming steps. Thus, the differentiation is based on cationic, anionic, radical, pericyclic, photochemical, transition metal-catalyzed, oxidative or reductive, and enzymatic reactions.

An example of a domino process induced by an anionic transformation is the domino Knoevenagel/hetero-Diels–Alder reaction. This was applied for the total synthesis of the alkaloid hirsutine (**3**) [4], which was isolated from the plants

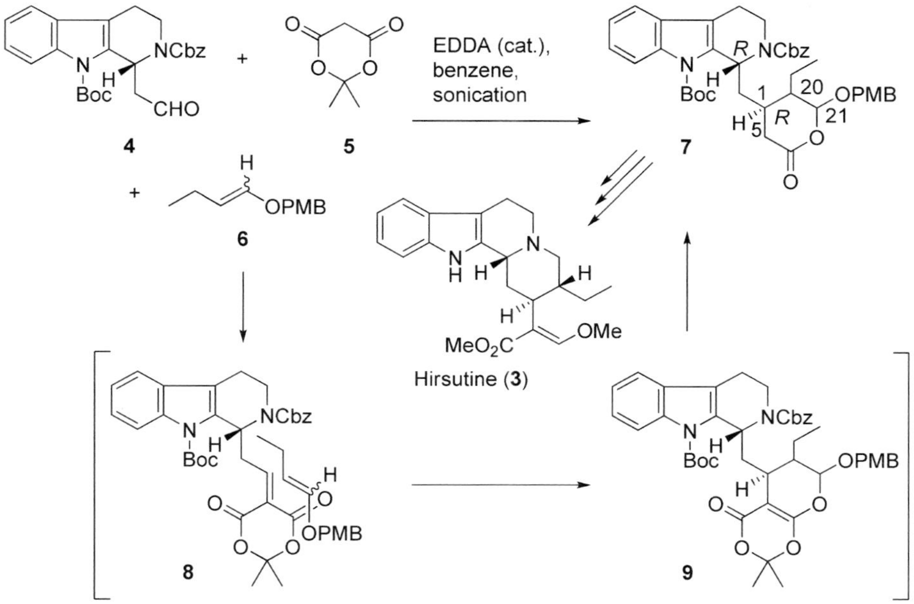

Scheme 2　Domino reaction for the synthesis of hirsutine (**3**).

Mitragyna hirsuta and *Uncaria rhynchophylla* MIQ [5]. The use of extracts of the latter plant has had a long tradition in the Chinese folk medicine »Kampo«. Hirsutine belongs to the corynanthe group of indole alkaloids and currently is attracting much attention because it shows a strong inhibition of the influenza A virus of the subtype H3N2 [6, 7].

In the course of the total synthesis of hirsutine (**3**), β-carboline **4**, which is obtained in enantiopure form from tryptamine over four steps [8], is reacted in a domino Knoevenagel/hetero-Diels–Alder transformation with Meldrum's acid (**5**) and an *E/Z* mixture of the enol ether **6** to give the lactone **7** [9].

Upon treatment of the mixture of **4**, **5** and **6** with catalytic amounts of ethylenediammonium diacetate (EDDA) in benzene, under sonication, the process proceeds as follows: in the first step the Knoevenagel condensation product **8** is formed from **4** and **5**, and this subsequently undergoes a hetero-Diels–Alder reaction with enol ether **6** to give the cycloadduct **9**. This intermediate is not stable under the reaction conditions and loses CO_2 and acetone by reaction with water formed in the condensation step to give the lactone **7** with an induced diastereoselectivity of >24:1 with reference to the stereogenic center C-15. The other two stereogenic centers are formed unselectively since the enol ether **6** is employed as a diastereomeric mixture. However, this is of no concern because these two stereogenic centers are lost during the further transformations leading to hirsutine (**3**) within five steps, again using a domino process which is induced by treatment of **7** with methanol and K_2CO_3 followed by a hydrogenation using Pd/C as catalyst. The transformation includes an opening of the lactone with formation of an aldehyde, deprotection of the secondary amine, and addition of the amine to the aldehyde with formation of an enamine, which is hydrogenated in a stereoselective way to give a single diastereomer.

The quality of a domino reaction is closely related to the number of bonds formed during the process and the increase in molecular complexity compared to the starting materials. While a plethora of two-bond-forming domino reactions have been reported, three-bond-forming transformations and higher remain the exception. One of the rare examples is the formation of the erythrina skeleton by a trimethylaluminum-mediated domino condensation type / iminium ion formation / iminium ion cyclization sequence [10]. The erythrina alkaloids such as erysodine (**10**) are a widespread class of natural products with extensive biological activity [11]. Many compounds of this family exhibit curare-like activity as well as CNS (central nerve system) depressant properties. For the synthesis, the primary amine **12** is first treated with $AlMe_3$ in benzene at

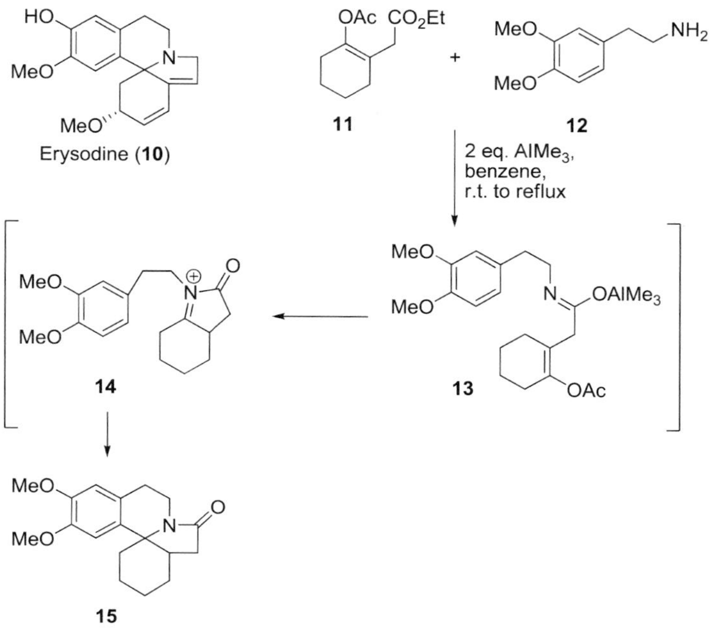

Scheme 3 Three-step domino reaction for the synthesis of erythrina alkaloids.

room temperature and, after addition of the enol acetate **11**, heated under reflux to give the erythrina scaffold **15**.

Mechanistic investigations using on-line NMR spectroscopy reveal that a metalated azaenolate **13** is formed first. An intramolecular addition to the enol acetate moiety in **13** with subsequent elimination of acetic acid then gives an *N*-acyliminium ion **14**. Finally, an electrophilic aromatic substitution leads to the erythrina skeleton **15**. Overall, three new bonds are formed selectively in this domino process, providing the alkaloid scaffold in very good yield (**79%**).

One of the first enantioselective transition metal-catalyzed domino processes has been employed for the total synthesis of vitamin E (**16**), which is the most important lipophilic antioxidant [12]. Thus, the reaction of phenol **17** with methyl vinyl ketone (**18**) in the presence of Pd(OTFA)₂, the chiral BOXAX ligand **19** and *p*-benzoquinone in dichloromethane at room temperature directly provides the chroman skeleton and part of the side chain of vitamin E [13]. The quarternary stereogenic center in the product **20** is formed with a high selectivity (97% *ee*) in the naturally occurring configuration, which is essential for the biological activity. Similar results are obtained using methyl acrylate (**27**) instead of methyl vinyl ketone.

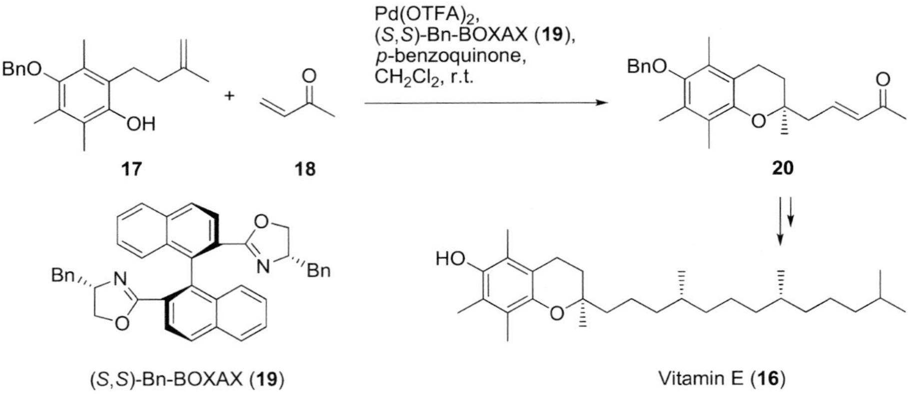

Scheme 4   Enantioselective domino reaction for the synthesis of vitamin E (**16**).

This domino transformation can be regarded as a combination of a Wacker type cyclization and a Heck reaction. It is assumed to start with the facially selective co-ordination of the chiral catalyst generated *in situ* from Pd(OTFA)$_2$ and the enantiomerically pure ligand (*S*,*S*)-Benzyl-BOXAX (**19**) to the olefinic double bond in **17** to form a π-complex **21**. Intramolecular oxypalladation provides the σ-complex **22** with the desired absolute configuration at the quarternary stereogenic center. Since β-hydride elimination is not possible, an intermolecular Heck reaction with methyl vinyl ketone (**18**) to form **23** occurs, and this can now undergo β-hydride elimination to release the product **20** and a Pd(0) species. To perform the reaction catalytically, the addition of *p*-benzoquinone as reoxidant is required to regenerate the active Pd(II) species.

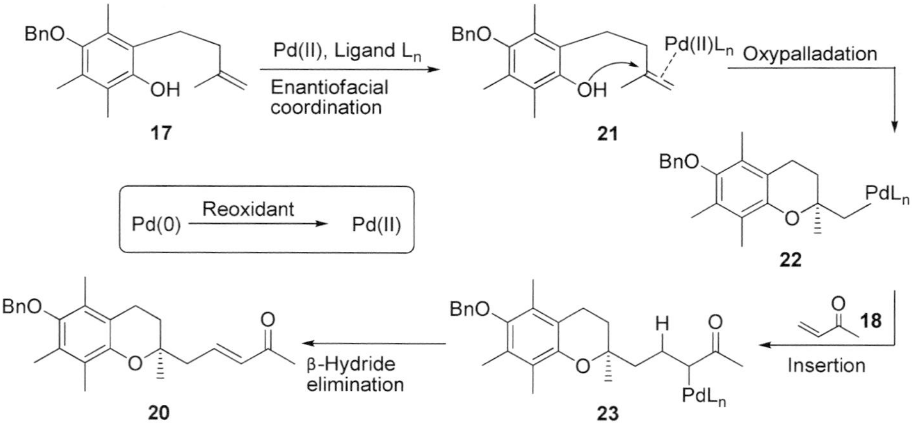

Scheme 5   Proposed mechanism of the domino Wacker–Heck reaction.

The domino Wacker–Heck reaction can also be performed omitting the chiral BOXAX ligand for an operationally simple synthesis of 2,3-dihydro-benzo[1,4]dioxins [14]. This two-step procedure, which is described, including experimental details, starts with the monoallylation of pyrocatechol (**24**) with 3-chloro-2-methylpropene (**25**). The obtained precursor **26** is subsequently employed in a domino Wacker–Heck reaction with methyl acrylate (**27**) to give the substituted 2,3-dihydro-benzo[1,4]dioxin *rac*-**28** as a racemic mixture (racemic since the reaction in the experimental part is performed in the absence of a chiral ligand) as a simplified version.

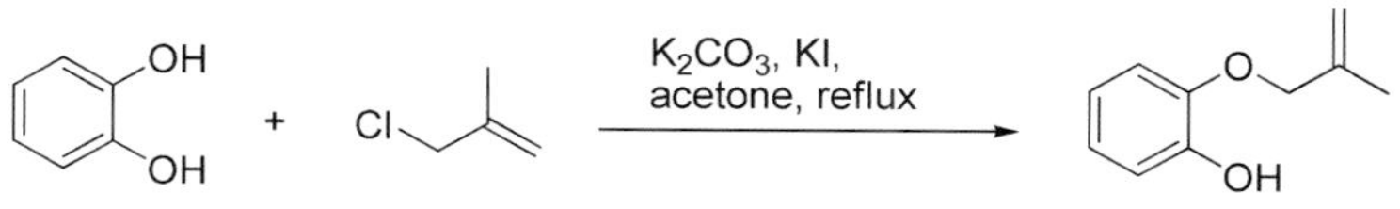

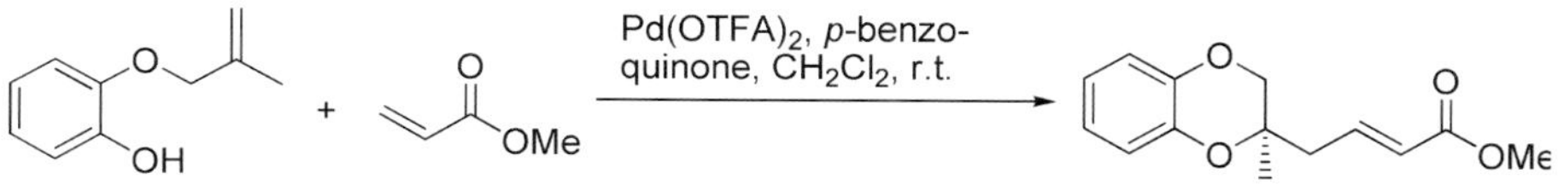

Scheme 6 Synthesis of 2,3-dihydrobenzo[1,4]dioxins.

Experimental details for a precursor and a domino reaction

Preparation of 2-(2-methylallyloxy)-phenol (26)

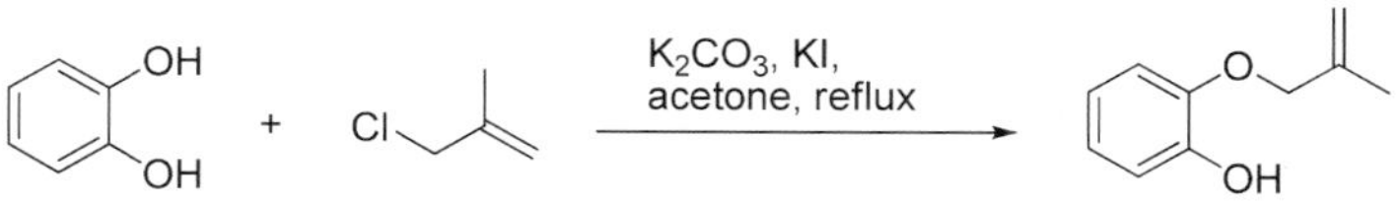

Apparatus   The following procedure is carried out under an argon atmosphere in a flame-dried 250 mL two necked round-bottomed flask equipped with a reflux condenser with an argon inlet and a magnetic stirring bar. Safety glasses, laboratory coat, and protective gloves must be worn.

Chemicals	Pyrocatechol, 3-chloro-2-methylpropene, K_2CO_3, KI, acetone.

Attention!   Wear suitable protective clothing, gloves and eye/face protection.

Caution!   Reaction should be carried out in a well-ventilated hood.

Pyrocatechol is harmful in contact with skin and if swallowed, and is irritating to the eyes. Do not breathe dust, and wear suitable gloves. In case of contact with eyes, rinse immediately with plenty of water and seek medical advice.

3-Chloro-2-methylpropene is highly flammable, corrosive, and dangerous for the enviroment. It is harmful by inhalation and if swallowed. It causes burns and may cause sensitization by skin contact. It is toxic to aquatic organisms and may cause long-term adverse effects in the aquatic environment. Keep container in a well-ventilated place and away from sources of ignition. No smoking. Do not empty into drains. In case of accident or if you feel unwell, seek medical advice immediately (show label where possible). Avoid release to the environment. Refer to special instructions safety data sheet.

Potassium carbonate (K_2CO_3) is harmful if swallowed and irritating to the eyes, respiratory system and skin. In case of contact with eyes, rinse immediately with plenty of water and seek medical advice.

Acetone is highly flammable and irritating to the eyes. Repeated exposure may cause skin dryness or cracking. Vapors may cause drowsiness and dizziness. Keep container in a well-ventilated place and away from sources of ignition. No smoking. In case of contact with eyes, rinse immediately with plenty of water and seek medical advice.

Potassium iodide (KI) has no health and safety hazard information indicated.

Procedure   A suspension of anhydrous K_2CO_3 (6.92 g, 50.1 mmol, 1.1 equiv.) and KI (8.30 g, 50.1 mmol, 1.1 equiv.) in acetone (80 mL) is treated with pyrocatechol (5.01 g, 45.5 mmol, 1.0 equiv.) and stirred at room temperature until the end of gas release (ca. 30 min). Afterwards 3-chloro-2-methylpropene (5.34 mL, 4.94 g, 54.6 mmol, 1.2 equiv.) is added and the resulting mixture stirred under reflux for 4 h. The mixture is cooled to room temperature and poured into water (200 mL). The solution is neutralized with HCl (1 mol L^{-1}), and the aqueous phase is extracted with diethyl ether (3 × 100 mL). The combined organic layers are dried over $MgSO_4$, the solvent evaporated *in vacuo* and the crude product purified by column chromatography on silica gel (*n*-pentane/ethyl acetate 50:1 → 10:1) to yield the title compound (3.90 g, 23.8 mmol, 52%) as a colorless liquid.

Characterization data: ^{1}H NMR (300 MHz, CDCl$_3$): δ 1.85 (s, 3H, 2'-CH$_3$), 4.51 (s, 2H, 1'-H$_2$), 5.03 (m$_c$, 1H, 3'-H), 5.10 (m$_c$, 1H, 3'-H), 5.70 (s, 1H, OH), 6.79-6.99 (m, 4H, Ar-H) ppm. ^{13}C NMR (75.6 MHz, CDCl$_3$): δ 19.37 (2'-CH$_3$), 72.59 (C-1'), 113.3 (C-3'), 112.1, 114.6, (C-3, C-6), 120.0, 121.6 (C-4, C-5), 140.4 (C-2'), 145.6, 145.8 (C-1, C-2) ppm. IR (film): $\bar{v}$ 3536, 2919, 1597, 1501, 1454, 1373, 1259, 1219 cm^{-1}. UV (MeCN): λ_{max} (lg ε) = 197.0 (4.582), 199.5 (4.582), 215.5 (3.798), 276.0 nm (3.424). MS (EI, 70 eV): m/z 164 ([M]$^+$, 57%), 109 ([M–C$_4$H$_7$]$^+$, 30%), 55 ([C$_4$H$_7$]$^+$, 100%).

Waste Disposal Solvent-free aqueous layers are disposed of to the sewer system. Organic solvents and compounds are collected.

Preparation of (*rac*)-(*E*)-4-(2-methyl-2,3-dihydrobenzo[b][1,4]dioxin-2-yl)-but-2-enoate (28)

26 + 27 → rac-28

Apparatus The following procedure is carried out under an argon atmosphere in a flame-dried 5 mL two-necked round-bottomed flask equipped with an argon inlet and a magnetic stirring bar. Safety glasses, laboratory coat, and protective gloves must be worn.

Chemicals Methyl acrylate, palladium(II) trifluoroacetate, *p*-benzoquinone, dichloromethane.

Attention! Wear suitable protective clothing, gloves and eye/face protection.

Caution! Reaction should be carried out in a well-ventilated hood.

Methyl acrylate is highly flammable and harmful by inhalation, in contact with skin and if swallowed. It is irritating to the eyes, respiratory system and skin (may cause sensitization by skin contact). Keep container in a well-ventilated place and avoid contact with eyes. In case of contact with eyes, rinse immediately with plenty of water and seek medical advice. Take precautionary measures against static discharges.

Palladium(II) trifluoroacetate is irritating to the eyes, respiratory system and skin. In case of contact with eyes, rinse immediately with plenty of water and seek medical advice.

p-Benzoquinone is toxic by inhalation and if swallowed. It is also irritating to the eyes, respiratory system and skin. It is very toxic to aquatic organisms. In case of contact with eyes, rinse immediately with plenty of water and seek medical advice. After contact with skin, wash immediately with plenty of water. In case of accident or if you feel unwell, seek medical advice immediately (show label where possible). Avoid release to the environment. Refer to special instructions safety data sheet.

Dichloromethane is harmful and there is limited evidence of a carcinogenic effect. Do not breathe fumes and avoid contact with skin and eyes.

Procedure

A mixture of palladium(II) trifluoroacetate (10.1 mg, 30.3 μmol, 0.1 equiv.) and *p*-benzoquinone (131 mg, 1.21 mmol, 4.0 equiv.) in CH_2Cl_2 (0.20 mL) is stirred for 10 min at room temperature. Then a solution of 2-(2-methylallyloxy)phenol (49.8 mg, 303 μmol, 1.0 equiv.) and methyl acrylate (55.0 μL, 52.2 mg, 607 μmol, 2.0 equiv.) in CH_2Cl_2 (0.20 mL) is added via a syringe to the suspension, and the mixture is stirred for 12 h at room temperature. At the end of the reaction the mixture is treated with 10 mL HCl (1 mol L^{-1}) and the aqueous phase extracted with diethyl ether (3 × 10 mL). The combined organic phases are washed with 3 × 10 mL NaOH (1 mol L^{-1}) and dried over $MgSO_4$, and the solvent is removed under reduced pressure. The crude product is purified by column chromatography on silica gel (*n*-pentane/ethyl acetate 6:1) to give the title compound (66.1 mg, 266 μmol, 88%) as a colorless oil.

Characterization data: ^1H NMR (300 MHz, CDCl$_3$): δ 1.32 (s, 3H, 2'-CH$_3$), 2.48 (ddd, J = 14.4, 8.3, 1.3 Hz, 1H, 4-H$_a$), 2.60 (ddd, J = 14.2, 7.2, 1.6 Hz, 1H, 4-H$_b$), 3.74 (s, 3H, OMe), 3.85 (d, J = 11.3 Hz, 1 H, 3'-H$_a$), 3.96 (d, J = 11.3 Hz, 1 H, 3'-H$_b$), 5.92 (dt, J = 15.7, 1.4 Hz, 1H, 2-H), 6.79-6.92 (m, 4H, Ar-H), 7.00 (ddd, J = 15.6, 8.2, 7.2 Hz, 1H, 3-H) ppm. ^{13}C NMR (75.6 MHz, CDCl$_3$): δ 21.23 (2'-CH$_3$),

38.48 (C-4), 51.54 (OMe), 70.28 (C-2″), 73.58 (C-3′), 117.0, 117.6 (C-5′, C-8′), 121.1, 122.0 (C-6′, C-7′), 124.8 (C-2), 142.0, 142.1 (C-4′a, C-8′a), 142.6 (C-3), 166.4 (C-2) ppm. IR (film): $\bar{v}$ 2981, 1724, 1659, 1594, 1494, 1436, 1264 cm^{-1}. UV (MeCN): λ_{max} (lg ε) = 201.1 (4.720), 277.5 nm (3.472). MS (EI, 70 eV): m/z 248 ([M]$^+$, 24%), 149 ([M–C$_5$H$_7$O$_2$]$^+$, 100%).

Waste Disposal Solvent-free aqueous layers are disposed of to the sewer system. Organic solvents and compounds are collected.

References and Notes

1 K.C. Nicolaou, D. Vourloumis, N. Winssinger, P.S. Baran, *Angew. Chem.* **2000**, *112*, 46; *Angew. Chem. Int. Ed.* **2000**, *39*, 44.
2 (a) L.F. Tietze, G. Brasche, K.M. Gericke, *Domino Reactions in Organic Synthesis*, Wiley-VCH, Weinheim, **2006**; (b) L.F. Tietze, *Chem. Rev.* **1996**, *96*, 115; (c) L.F. Tietze, U. Beifuß, *Angew. Chem.* **1993**, *105*, 137; *Angew. Chem. Int. Ed.* **1993**, *32*, 131.
3 K.U. Wendt, G.E. Schulz, E.J. Corey, D.R. Liu, *Angew. Chem.* **2000**, *112*, 2930; *Angew. Chem. Int. Ed.* **2000**, *39*, 2812.
4 The name hirsutine was also used for a sesquiterpene of the triquinone type. Nowadays, these compounds are named as hirsutane and hirsutene.
5 (a) E.J. Shellard, A.H. Becket, P. Tantivatana, J.D. Phillipson, C.M. Lee, *J. Pharm. Pharmacol.* **1966**, *18*, 553; (b) G. Laus, H. Teppner, *Phyton* **1996**, *36*, 185.
6 H. Takayama, Y. Limura, M. Kitajima, N. Aimi, K. Konno, *Bioorg. Med. Chem. Lett.* **1997**, *7*, 3145.
7 L.F. Tietze, A. Modi, *Med. Res. Rev.* **2000**, *20*, 304.
8 L.F. Tietze, Y. Zhou, E. Töpken, *Eur. J. Org. Chem.* **2000**, 2247.
9 L.F. Tietze, Y. Zhou, *Angew. Chem.* **1999**, *111*, 2076; *Angew. Chem. Int. Ed.* **1999**, *38*, 2045.
10 S.A.A. El Bialy, H. Braun, L.F. Tietze, *Angew. Chem.* **2004**, *116*, 5505; *Angew. Chem. Int. Ed.* **2004**, *43*, 5391.
11 (a) A.S. Chawla, V.K. Kapoor, in: *Handbook of Plant and Fungal Toxicants* (Ed.: J.F. D'Mello), CRC-Press, London, **1997**, pp. 37–49; (b) D.S. Bhakuni, *J. Ind. Chem. Soc.* **2002**, *79*, 203.
12 (a) T. Netscher, *Chimia* **1996**, *50*, 563–567; (b) M. K. Horwitt, *Am. J. Clin. Nutr.* **1986**, *44*, 973; (c) H.M. Evans, K.S. Bishop, *Science* **1922**, *56*, 650.
13 (a) L.F. Tietze, K.M. Sommer, J. Zinngrebe, F. Stecker, *Angew. Chem.* **2005**, *117*, 262; *Angew. Chem. Int. Ed.* **2005**, *44*, 257; (b) L.F. Tietze, F. Stecker, J. Zinngrebe, K.M. Sommer, *Chem. Eur. J.* **2006**, *12*, 8770.
14 L.F. Tietze, K.F. Wilckens, S. Yilmaz, F. Stecker, J. Zinngrebe, *Heterocycles* **2006**, *70*, 309.

25

N-Heterocyclic Phosphenium Salts

Charles L. B. Macdonald, Bobby D. Ellis, Gregor Reeske and Alan H. Cowley

The study of compounds containing main-group elements in unusual coordination environments has been a central aspect of research into the chemistry of the s- and p-block elements, and has been crucial to the better understanding of the nature of structure and bonding of inorganic, organometallic and even organic molecules. Phosphorus compounds containing a cationic two-coordinate phosphorus center having only 6 valence electrons are known as »phosphenium« cations and are generally very reactive and electrophilic species – isovalent with carbenes and silylenes – that exhibit a rich and interesting chemistry [1–3].

The diverse chemistry of a class of remarkably stable carbenes known as N-heterocyclic carbenes (NHCs) has had a tremendous impact on synthetic chemistry and catalysis since the early 1990s [4, 5]. The importance of NHCs has spurred research into the isovalent analogs containing other main-group elements; the phosphorus analogs may be called N-heterocyclic phosphenium cations (NHPs). The chemistry of NHPs has proven to be quite distinct from that of NHCs, and the unique reactivity of such compounds is currently being exploited [6–13]. However, the synthesis of such compounds has not been efficient [14]. As a solution to this problem, our groups have developed simple, quantitative and remarkably clean one-step redox routes to N-heterocyclic phosphenium cations [15–17].

Preparation of N-heterocyclic phosphenium salts (2 stages)

Preparation of [(DAB)P][I$_3$], DAB = [(2,6-i Pr$_2$C$_6$H$_3$)N=CMe]$_2$

Apparatus   Two 100 mL one-necked Schlenk flasks, each equipped with a rubber septum and a magnetic stirring bar, one syringe (50 mL), one cannula, one needle, safety glasses, protective gloves, a Schlenk manifold, nitrogen- or argon-filled dry box (or glovebag). Safety glasses, laboratory coat and protective gloves.

Chemicals DAB, PI$_3$, dry dichloromethane, dry acetonitrile (optional).

Attention! Safety glasses and protective gloves must be worn at all times.

Caution! Because of their toxicity and volatility, care should be taken to avoid inhalation of dichloromethane and phosphorus triiodide or contact of their solutions with the skin. All reactions should be carried out in a well-ventilated hood.

Procedure In an inert atmosphere (dry box or glovebag), one of the oven-dried Schlenk flasks is charged with a magnetic stirring bar and DAB (0.352 g, 0.870 mmol), and the other with a magnetic stirring bar and PI$_3$ (0.358 g, 0.870 mmol); both flasks are then sealed with rubber septa. On a Schlenk line, each of the reagents is then dissolved with magnetic stirring in CH$_2$Cl$_2$ (15 mL) added by cannula or syringe. At room temperature, the solution containing DAB is then added drop by drop by cannula to the solution containing PI$_3$ to produce a dark red solution. The solution is left to stir overnight, and the volatile components are then removed under reduced pressure to give [(DAB)P][I$_3$] in quantitative yield as a red solid. Dissolution of the red solid in a minimal amount of MeCN followed by slow evaporation of the solvent in a glovebox provides red crystalline material. Yield: 84% (0.597 g, 0.731 mmol).

Characterization data: m.p. 290–291 °C. ^{31}P{^{1}H} NMR (CD$_2$Cl$_2$): δ 201.1 (s) ppm. ^{13}C{^{1}H} NMR (CD$_2$Cl$_2$): δ 13.9(s), 23.3 (s), 26.1(s), 29.5 (s), 125.0 (s), 125.8 (s), 132.9 (s), 144.3 (s), 145.5 (s) ppm. ^{1}H NMR (CD$_2$Cl$_2$): δ 1.22 (d, J = 7.0 Hz, 24H), 2.31 (s, 6H), 2.67 (septet, J = 7.0 Hz, 2H), 7.48 (d, J = 7.9 Hz, 4H), 7.67 (t, J = 7.9 Hz, 2H) ppm. Calculated (found) for C$_{28}$H$_{40}$I$_3$N$_2$P: C 41.20 (41.47), H 4.94 (4.93)%.

Waste Disposal If the evaporated CH$_2$Cl$_2$ is collected in a trap during the isolation of the salt, it may be distilled and reused; otherwise, it should be disposed of in an appropriate container for chlorinated waste.

Preparation of [(DAB)P][SnCl$_5$], DAB = [(2,6-iPr$_2$C$_6$H$_3$)N=CMe]$_2$

Apparatus

Two 100 mL one necked Schlenk flasks, each equipped with a rubber septum and a magnetic stirring bar, one syringe (50 mL), one cannula, one needle, safety glasses, protective gloves, a Schlenk manifold, nitrogen- or argon-filled dry box or glovebag, fritted glass filter with Celite (optional), safety glasses, laboratory coat and protective gloves.

Chemicals

DAB, anhydrous SnCl$_2$, dry PCl$_3$, dry dichloromethane.

Attention!

Safety glasses and protective gloves must be worn at all times.

Caution!

Because of their toxicity, corrosiveness (for PCl$_3$), and volatility, care should be taken to avoid inhalation of dichloromethane and phosphorus trichloride or contact of their solutions with the skin. All reactions should be carried out in a well-ventilated hood.

Procedure

In an inert atmosphere, one of the oven-dried Schlenk flasks is loaded with PCl$_3$ (0.147 g, 93.4 µL, 1.070 mmol) and sealed with a rubber septum. The other flask is charged with DAB (0.433 g, 1.070 mmol), SnCl$_2$ (0.203 g; 1.070 mmol) and a magnetic stirring bar and then sealed with a rubber septum. On the Schlenk line, CH$_2$Cl$_2$ (20 mL) is added by cannula or syringe to the DAB/SnCl$_2$ mixture, with stirring, to produce a yellow slurry. A similar amount of CH$_2$Cl$_2$ (20 mL) is added to the flask containing PCl$_3$, and the resultant solution is then added drop by drop by cannula to the stirred slurry of DAB and SnCl$_2$. The mixture becomes red and is left to stir overnight. If any undissolved material remains at this time, the mixture may be filtered through a plug of Celite. Slow evaporation of the solution in the dry box produces pink crystalline material. Yield: 88% (0.692 g, 0.946 mmol).

Characterization data: m.p. 215–216 °C. ^{31}P{^{1}H} NMR (CD$_2$Cl$_2$): δ 200.7 (s) ppm. ^{13}C{^{1}H} NMR (CD$_2$Cl$_2$): δ 13.5 (s), 23.2 (s), 26.1(s), 29.4(s), 124.9(s), 125.8(s), 132.9(s), 144.3(s), 145.5(s) ppm. ^{1}H NMR (CD$_2$Cl$_2$): δ 1.21 (d, $J = 6.9$, 24H), 2.31 (s, 6H), 2.67 (septet, $J = 6.9$ Hz, 2H), 7.49 (d, $J = 7.8$ Hz, 4H), 7.69 (t, $J = 7.8$ Hz, 2H) ppm. Calculated (found) for C$_{28}$H$_{40}$Cl$_5$N$_2$PSn: C 45.97 (45.78), H 5.51(5.45)%.

Waste Disposal If the evaporated CH_2Cl_2 is collected in a cold trap during the isolation of the salt, it should be disposed of in an appropriate container for chlorinated waste.

Explanation

Cyclo-addition reactions are among the cleanest types of reactions because many of them proceed spontaneously, in very high yield, and without the formation of any by-products; some of the cleanest examples are the organic reactions described by the term »click chemistry« [18]. The cleanliness of the strategies for the formation of the NHP salts described in reactions (1) and (2) below may be understood in the context of cyclo-addition chemistry and the redox chemistry of putative P(+1) cations (Scheme 1).[19]

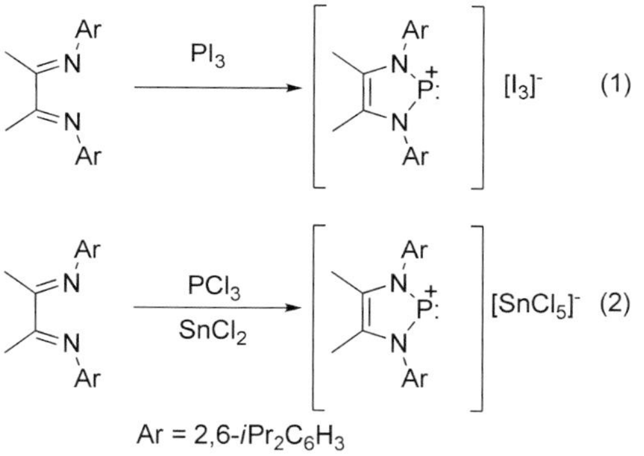

Scheme 1 NHP salt formation, where Ar = 2,6-iPr$_2$C$_6$H$_3$.

It has been demonstrated that, in the presence of appropriate ligands, either PI$_3$ or the mixture of PCl$_3$ with SnCl$_2$ generates »P(+1)« cations. The quantitative production of the P(+3)-containing NHP salts from the P(+1) synthetic strategies employs »click«-like cyclo-addition reactivity to form the ring and involves the formal transfer of 2 electrons from the phosphorus center to the DAB ligand (Scheme 2).

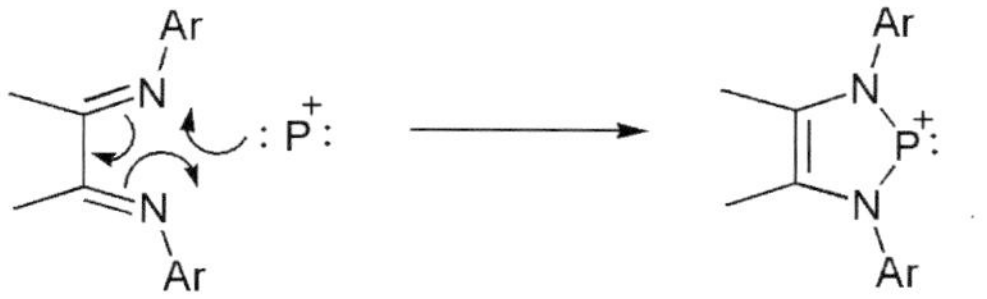

Scheme 2 Electron transfer from phosphorus, where Ar = 2,6-iPr$_2$C$_6$H$_3$.

The absence of by-products from the redox syntheses described above is readily understood because, in each case, the products formed by the redox reaction used to generate the P(+1) center are incorporated into the anion of each salt: the I_2 produced in the first reaction becomes part of the $[I_3]^-$ anion and the $SnCl_4$ formed in the second reaction is captured in the $[SnCl_5]^-$ anion.

In the context of sustainable chemistry, it is also worthy of mention that the α-diimine DAB ligand is itself synthesized in a condensation reaction of an α-diketone with a primary amine that involves only the evolution of water as a by-product [20]. Furthermore, it should be noted that many other α-diimine ligands have proven to be amenable to the synthetic approaches described above [16] and that the anion $[SnCl_5 \cdot thf]^-$ has been observed when the pentachlorostannate salt is crystallized from THF [15]. For the purposes of further synthetic utility, it should also be mentioned that the triiodide salts can be turned into salts containing other anions using salt metathesis chemistry [16].

References

1 A.H. Cowley, R. A. Kemp, *Chem. Rev.* **1985**, *85*, 367.
2 M. Sanchez, M.-R. Mazieres, L. Lamande, R. Wolf, in *Multiple Bonds and Low Coordination in Phosphorus Chemistry* (Eds.: M. Regitz, O. J. Scherer), Thieme Verlag, Stuttgart, **1990**, pp. 129.
3 D. Gudat, *Coord. Chem. Rev.* **1997**, *163*, 71.
4 A. J. Arduengo, *Acc. Chem. Res.* **1999**, *32*, 913.
5 D. Bourissou, O. Guerret, F. P. Gabbai, G. Bertrand, *Chem. Rev.* **2000**, *100*, 39.
6 D. Gudat, A. Haghverdi, H. Hupfer, M. Nieger, *Chem. Eur. J.* **2000**, *6*, 3414.
7 D. Gudat, A. Haghverdi, M. Nieger, *J. Organomet. Chem.* **2001**, *617*, 383.
8 D. Gudat, A. Haghverdi, T. Gans-Eichler, M. Nieger, *Phosphorus Sulfur Silicon Relat. Elem.* **2002**, *177*, 1637.
9 S. Burck, D. Gudat, M. Nieger, *Angew. Chem.* **2004**, *116*, 4905; *Angew. Chem. Int. Ed.* **2004**, *43*, 4801.
10 S. Burck, D. Gudat, F. Lissner, K. Nättinen, M. Nieger, T. Schleid, *Z. Anorg. Allg. Chem.* **2005**, *631*, 2738.
11 S. Burck, D. Gudat, M. Nieger, W. W. Du Mont, *J. Am. Chem. Soc.* **2006**, *128*, 3946.
12 S. Burck, A. Daniels, T. Gans-Eichler, D. Gudat, K. Nättinen, M. Nieger, *Z. Anorg. Allg. Chem.* **2005**, *631*, 1403.
13 S. Burck, D. Förster, D. Gudat, *Chem. Commun.* **2006**, 2810.
14 S. Burck, D. Gudat, K. Nättinen, M. Nieger, M. Niemeyer, D. Schmid, *Eur. J. Inorg. Chem.* **2007**, 5112.
15 G. Reeske, C. R. Hoberg, N. J. Hill, A. H. Cowley, *J. Am. Chem. Soc.* **2006**, *128*, 2800.
16 G. Reeske, A. H. Cowley, *Inorg. Chem.* **2007**, *46*, 1426.
17 B. D. Ellis, C. L. B. Macdonald, *Inorg. Chim. Acta* **2007**, *360*, 329.
18 H. C. Kolb, M. G. Finn, K. B. Sharpless, *Angew. Chem. Int. Ed.* **2001**, *40*, 2004; *Angew. Chem.* **2001**, *113*, 2056.
19 B. D. Ellis, C. L. B. Macdonald, *Coord. Chem. Rev.* **2007**, *251*, 936.
20 H. tom Dieck, M. Svoboda, T. Greiser, *Z. Naturforsch.,B: Anorg. Chem., Org. Chem.* **1981**, *36B*, 823.

26

Methyltitanium Triisopropoxide

Armin de Meijere and Alexander Lygin

Early experiments concerning the synthesis of organotitanium compounds date back to 1861, when Cahours reported on some reactions involving $TiCl_4$ and $ZnEt_2$ [1]. However, since the sensitive nature of organotitanium compounds had not been anticipated during these and other early experiments, they generally just led to intractable residues. Improved procedures for handling air-sensitive and thermally unstable compounds enabled chemists to synthesize and fully characterize the first organotitanium compound, $PhTi(OiPr)_3$, only in 1952 [2]. In 1955 Ziegler et al. first reported [3] that the dark solid product, formed upon reaction of alkylaluminum compounds with titanium halides, is an active catalyst for the polymerization of ethylene at ambient pressure. This discovery had a great influence on the subsequent development of science and technology and encouraged scientists to further explore organotitanium compounds. Various such reagents were employed as catalysts or mediators of different types for organic transformations [4], including first of all the Ziegler-Natta alkene polymerization [5], but also the Sharpless asymmetric epoxidation [6], the McMurry reductive coupling of carbonyl compounds [7] and others. Certain titanacyclopropanes (titanium alkene complexes), formed *in situ* from titanium tetraisopropoxide or chlorotitanium triisopropoxide and alkylmagnesium halides, bring about the transformation of alkyl carboxylates to cyclopropanols, *N,N*-dialkylcarboxamides to *N,N*-dialkylcyclopropylamines [8], and nitriles to primary cyclopropylamines [9].

We found that methyltitanium triisopropoxide, easily prepared from titanium tetraisopropoxide, titanium tetrachloride and methyllithium [10], chlorotitanium triisopropoxide and methylmagnesium bromide [11], or *in situ* from $Ti(OiPr)_4$ and methylmagnesium chloride [12], more efficiently mediates the reductive cyclopropanation of *N,N*-dialkylcarboxamides [12, 13].

Two reactions mediated by methyltitanium triisopropoxide

Preparation of *N*-cyclopropylmorpholine

Apparatus   100 mL Schlenk flask with a septum, a magnetic stir bar, and a T-shaped N_2 inlet; syringe (10 mL); safety glasses; protective gloves.

Chemicals   MeTi(O*i*Pr)$_3$, freshly prepared EtMgBr (3.35 M solution in Et$_2$O), anhydrous tetrahydrofuran, *N*-formylmorpholine.

Attention!   Safety glasses and protective gloves must be worn at all times.

Caution!   MeTi(O*i*Pr)$_3$ and EtMgBr both undergo exothermic hydrolysis in the presence of water. Care should also be taken to avoid inhalation of tetrahydrofuran, diethyl ether or dichloromethane, or contact of such solutions with the skin, because of their toxicity. All operations should be carried out in a well-ventilated hood.

Procedure   To a solution of methyltitanium triisopropoxide (2.88 g, 12 mmol) and *N*-formylmorpholine (1.15 g, 10 mmol) in anhydrous tetrahydrofuran (48 mL), kept in an oven-dried 100 mL Schlenk flask under an atmosphere of dry nitrogen, is added drop by drop with stirring a 3.35 M solution of EtMgBr in Et$_2$O (6.0 mL, 20 mmol) at r.t. over a period of 10 min. The mixture is stirred at r.t. for 16 h before the reaction is quenched by addition of water (10 mL). The mixture is exposed to air, until the precipitate turns colorless (~20 min). It is then removed by filtration through a pad of Celite, and the latter is washed with dichloromethane (5 × 30 mL). The filtrate is treated once with a saturated solution of NaHCO$_3$ (30 mL) and dried over anhydrous Na$_2$SO$_4$. The solvents are removed under reduced pressure to give a crude product, which is purified by column chromatography on silica gel (hexane/diethyl ether 1 : 1, $R_f = 0.28$) to give *N*-cyclopropylmorpholine (1.08 g, 85%) as a light yellow oil.

Characterization data: ^{1}H NMR (300 MHz, CDCl$_3$): δ 3.68 (s, 4H, CH$_2$), 2.61 (s, 4H, CH$_2$), 1.69–1.60 (m, 1H, CH), 0.48–0.38 (m, 4H, CH$_2$-cPr) ppm. ^{13}C NMR (75.5 MHz, CDCl$_3$): δ 67.1 (2CH$_2$), 53.9 (2CH$_2$), 39.0 (CH), 5.7 (2CH$_2$–cPr) ppm. IR (KBr, film, cm^{-1}): $\bar{v}$ 2959, 2855, 2808, 1740, 1452, 1362, 1265, 1214, 1120, 1016, 990, 881, 734. MS (DCI): *m/z* 128 (M+H$^+$, 100%). Calculated (found) for C$_7$H$_{13}$NO: C, 66.10 (66.26); H, 10.30 (10.15); N, 11.01 (11.21)%.

Preparation of *N*-benzyl-*N'*-cyclopropylpiperazine

Apparatus 25 mL Schlenk flask with a septum, a magnetic stir bar, and a T-shaped N_2 inlet, syringe (10 mL), safety glasses, protective gloves.

Chemicals MeTi(O*i*Pr)$_3$, freshly prepared EtMgBr (3.35 M solution in Et$_2$O), anhydrous tetrahydrofuran, *N*-benzyl-*N'*-formylpiperazine.

Attention! Safety glasses and protective gloves must be worn at all times.

Caution! MeTi(O*i*Pr)$_3$ and EtMgBr both undergo exothermic hydrolysis in the presence of water. Care should also be taken to avoid inhalation of tetrahydrofuran, diethyl ether, ethylacetate, or contact of such solutions with the skin because of their toxicity. All procedures should be carried out in a well-ventilated hood.

Procedure To a solution of methyltitanium triisopropoxide (900 mg, 3.75 mmol) and *N*-benzyl-*N'*-formylpiperazine (637 mg, 3.12 mmol) in anhydrous tetrahydrofuran (12 mL), kept in an oven-dried 25 mL Schlenk flask under an atmosphere of dry nitrogen, is added drop by drop with stirring a 3.35 M solution of EtMgBr in Et$_2$O (1.86 mL, 6.25 mmol) at r.t. over a period of 10 min. The mixture is stirred at r.t. for 14 h before the reaction is quenched by addition of water (5 mL). The mixture is exposed to air until the precipitate turns colorless (~15 min). It is then removed by filtration through a pad of Celite, and the latter is washed with ethyl acetate (5 × 20 mL). The filtrate is treated once with a saturated solution of NaHCO$_3$ (20 mL) and dried over anhydrous Na$_2$SO$_4$. The solvents are removed under reduced pressure to give a crude product, which is purified by column chromatography on silica gel (hexane/diethyl ether 1 : 1, R_f = 0.44) to give 623 mg (92%) of *N*-benzyl-*N'*-cyclopropylpiperazine as a light yellow oil.

Characterization data: ^{1}H NMR (300 MHz, CDCl$_3$): δ 7.32–7.20 (m, 5H, Ph), 3.51 (s, 2H, CH$_2$Ph), 2.65 (br s, 4H, 2CH$_2$N), 2.45 (br s, 4H, 2CH$_2$N), 1.64–1.56 (m, 1H, CH), 0.47–0.35 (m, 4H, 2CH$_2$-cPr) ppm. ^{13}C NMR (75.5 MHz, CDCl$_3$): δ 138.1 (C$_{quat}$, C$_{ipso}$), 129.2 (2CH, Ph), 128.1 (2CH, Ph), 127.0 (CH, Ph), 63.1 (CH$_2$Ph), 53.3 (2CH$_2$N), 52.9 (2CH$_2$N), 38.4 (CH), 5.66 (2CH$_2$-cPr) ppm. IR (KBr, film, cm^{-1}): $\bar{v}$ 2935, 2808, 1452, 1362, 1265, 1214, 1016, 881, 734, 709. MS (70eV): *m/z* 216 (M$^+$, 60%), 146 (M-70, 100%), 91 (PhCH$_2$, 52%). Calculated (found) for $C_{14}H_{20}N_2$: C, 77.73 (77.83); H, 9.32 (9.26); N, 12.95 (13.05)%.

Waste Disposal All organic solvents used in the preparation and isolation of the products are collected for proper disposal. Aqueous phases are neutralized before admitting them to the sewage. Solid wastes containing titanium dioxide, magnesium hydroxide, Celite, and used silica gel are collected separately for admission to the appropriate solid waste disposal system.

Explanation

Because of the unique steric and electronic properties of the cyclopropyl group [15] in conjunction with its high metabolic stability, more than 200 compounds containing a cyclopropylamine have found their relevance in medicinal chemistry [16–20], and many of them are N-cyclopropylated nitrogen heterocycles. Four conceptually different methods for the attachment of a cyclopropyl moiety to a nitrogen atom in a heterocycle have been developed [16, 21–23]. Among them, the reductive amination of cyclopropanone hemiacetal [16], the utilization of cyclopropylidene magnesium intermediates [21], copper-catalyzed cyclopropylation with tricyclopropyl bismuth [22] and copper-mediated coupling with cyclopropylboronic acid [23] require rather expensive reagents. Our previously reported [12–14] reductive cyclopropanation of N,N-dialkylcarboxamides with titaniumalkene complexes (titanacyclopropanes), *in situ* generated from alkylmagnesium halides and Ti(OiPr)₄ [14] or even better MeTi(OiPr)₃ [12, 13], can be applied to certain N-formyl-substituted nitrogen heterocycles, and thus provides a more efficient method to prepare such N-cyclopropyl-nitrogen heterocycles.

MeTi(OiPr)₃ is a commercially available reagent, but can also easily be prepared according to literature procedures [10, 11]. MeTi(OiPr)₃ has also been found to be superior to Ti(OiPr)₄ and ClTi(OiPr)₃ in the conversion of carboxylic acid esters to cyclopropanols [24].

References

1 (a) A. Cahours, *Ann. Chim. Phys.* **1861**, *62*, 257; (b) A. Cahours, *Liebigs Ann. Chem.* **1862**, *122*, 48.

2 D.F. Herman, W.K. Nelson, *J. Am. Chem. Soc.* **1952**, *74*, 2693.

3 K. Ziegler, E. Holzkamp, H. Breil, H. Martin, *Angew. Chem.* **1955**, *67*, 541.

4 (a) M.T. Reetz, *Organotitanium Reagents in Organic Synthesis*, Springer: Berlin/Heidelberg, **1986**; (b) M.T. Reetz, *Top. Curr. Chem.* **1982**, *106*, 1.

5 (a) H.G. Alt, A. Köppl, *Chem. Rev.* **2000**, *100*, 1205; (b) A.L. McKnight, R.M. Waymouth, *Chem. Rev.* **1998**, *98*, 2587.

6 Q.-H. Xia, H.-Q. Ge, C.-P. Ye, Z.-M. Liu, K.-X. Su, *Chem. Rev.* **2005**, *105*, 1603.

7 J.E. McMurry, *Acc. Chem. Res.* **1983**, *16*, 405.

8 Reviews, see: (a) O.G. Kulinkovich, A. de Meijere, *Chem. Rev.* **2000**, *100*, 2789; (b) A. de Meijere, S.I. Kozhushkov, A.I. Savchenko, *J. Organomet. Chem.* **2004**, *689*, 2033; (c) A. de Meijere, S.I. Kozhushkov, A.I. Savchenko in: *Titanium and Zirconium in Organic Synthesis* (Ed.: I. Marek), Wiley-VCH, Weinheim **2002**, 390.

9 P. Bertus, J. Szymoniak, *Chem. Commun.* **2001**, 1792.

10 A 500-mL, three-necked, round-bottomed flask equipped with two rubber septa and a dropping funnel fitted with an argon inlet is charged with a mixture of 44 mL (150 mmol) of titanium tetraisopropoxide in 50 mL of anhydrous ether and cooled to 0 °C in an ice bath. Titanium tetrachloride (5.48 mL, 50 mmol) is added drop by drop with a syringe pump over 30 min. The resulting mixture is allowed to warm to room temperature and stirred for 2 h. The reaction mixture is then cooled to 0 °C, and 123 mL (197 mmol) of a 1.6 M solution of methyllithium in ether is added from the dropping funnel over 40 min. The mixture is allowed to warm to room temperature and stirred for 1 h. The dropping funnel is replaced with a short-path distillation head, and the ether is removed by distillation at ambient pressure. Distillation of the crude product at reduced pressure (bp. 56-57 °C, 60 mm) gives 38.5 g (81%) of MeTi(OiPr)$_3$ as a yellow oil. See: A. de Meijere, H. Winsel, B. Stecker, *Org. Synth.* **2005**, *81*, 14.

11 H. Takahashi, K. Tanahashi, K. Higashiyama, H. Onishi, *Chem. Pharm. Bull.* **1986**, *34*, 479.

12 A. de Meijere, C.M. Williams, A. Kourdioukov, S.V. Sviridov, V. Chaplinski, M. Kordes, A.I. Savchenko, C. Stratmann, M. Noltemeyer, *Chem. Eur. J.* **2002**, *8*, 3789.

13 V. Chaplinski, H. Winsel, M. Kordes, A. de Meijere, *Synlett* **1997**, 111.

14 V. Chaplinski, A. de Meijere, *Angew. Chem.* **1996**, *108*, 491; *Angew. Chem. Int. Ed.* **1996**, *35*, 413.

15 (a) A. de Meijere, *Angew. Chem.* **1979**, *91*, 867; *Angew. Chem. Int. Ed.* **1979**, *18*, 809; (b) *The Chemistry of the Cyclopropyl Group, Parts 1,2* (Eds: S. Patai, Z. Rappoport) John Wiley & Sons: New York, **1987**; (c) *Houben-Weyl, Methods of Organic Chemistry*, Vol. E17, (Ed.: A. de Meijere) *Carbocyclic Three-Membered Ring Compounds*, Thieme, Stuttgart **1996**, 1.

16 M.L. Gillaspy, B.A. Lefker, W.A. Hada, D.J. Hoover, *Tetrahedron Lett.* **1995**, *36*, 7399.

17 J. Salaün, M.S. Baird, *Curr. Med. Chem.* **1995**, *2*, 511.

18 J. Salaün, *Top. Curr. Chem.* **2000**, *207*, 1.

19 L.A. Wessjohann, W. Brandt, T. Thiemann, *Chem. Rev.* **2003**, *103*, 1625.

20 (a) F. Brackmann, A. de Meijere, *Chem. Rev.* **2007**, *107*, 4493; (b) F. Brackmann, A. de Meijere, *Chem. Rev.* **2007**, *107*, 4538.

21 T. Satoh, M. Miura, K. Sakai, Y. Yokoyama, *Tetrahedron* **2006**, *62*, 4253.

22 A. Gagnon, M. St-Onge, K. Little, M. Duplessis, F. Barabe, *J. Am. Chem. Soc.* **2007**, *129*, 44.

23 T. Tsuritani, N.A. Strotman, Y. Yamamoto, M. Kawasaki, N. Yasuda, T. Mase, *Org. Lett.* **2008**, *10*, 1653.

24 J.C. Lee, M.J. Sung, J.K. Cha, *Tetrahedron Lett.* **2001**, *42*, 2059.

27

Ferrocenyl-1,2,3-Triazolyl Dendrimers by ›Click‹ Chemistry

Catia Ornelas, Jaime Ruiz and Didier Astruc

The long-known Huisgen 1,3-dipolar cycloaddition of terminal alkynes with primary, secondary and tertiary organic azides yielding triazoles is certainly one of the most useful cycloaddition reactions [1, 2]. It has been largely improved a few years ago by the research groups of Sharpless [3, 4] and Tonoe [5]. The term »click chemistry« was coined by Sharpless for this reaction and a few others [3, 6] under the new »green«[7] reaction conditions that make them easy to carry out and attractive. These improvements include catalysis by Cu(I) that is conveniently generated from a Cu(II) salt and sodium ascorbate, rendering the reaction regioselective in 1,4-disubstituted 1,2,3-triazole (Scheme 1). A variety of solvents can be used, such as *tert*-butyl alcohol and aqueous media such as water/THF or water/alcohol under rather mild reaction conditions (20–60 °C). Moreover, it does not need protection from air and tolerates a large variety of functional groups.

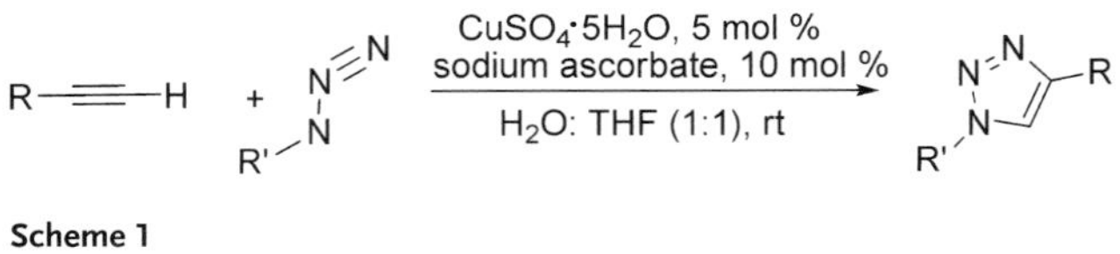

Scheme 1

Given the easy synthesis of organic azides from organic halides and sodium azide and the availability of terminal alkynes, this »click« reaction immediately became very popular and generally used to link together two molecular fragments with applications in biochemistry, materials chemistry and other purposes. This reaction was utilized for instance to synthesize dendrimers including their multi-generation construction [8, 9] (Scheme 2 below: R = dendron) [10] or the introduction of specific groups such as ferrocene [10–13] and sulfonate [14] at their periphery.

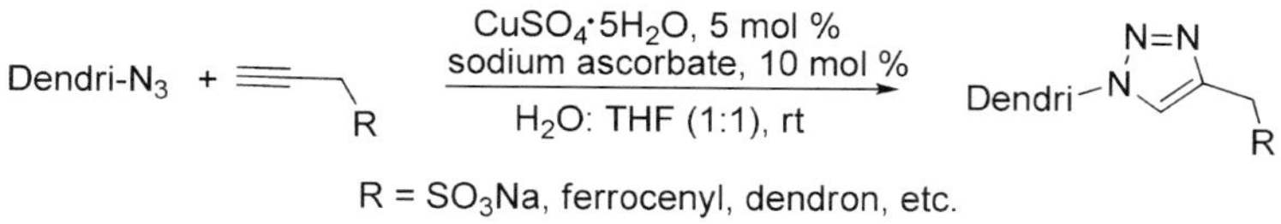

Scheme 2

Although the click reaction is catalytic in Cu(I), we found that the catalyst remained trapped inside the dendrimer so that completion could only be reached with a stoichiometric amount of Cu(I) [10, 12]. Here we describe the experimental procedure for the synthesis of a nona-branched click ferrocenyl dendrimer (Scheme 3).

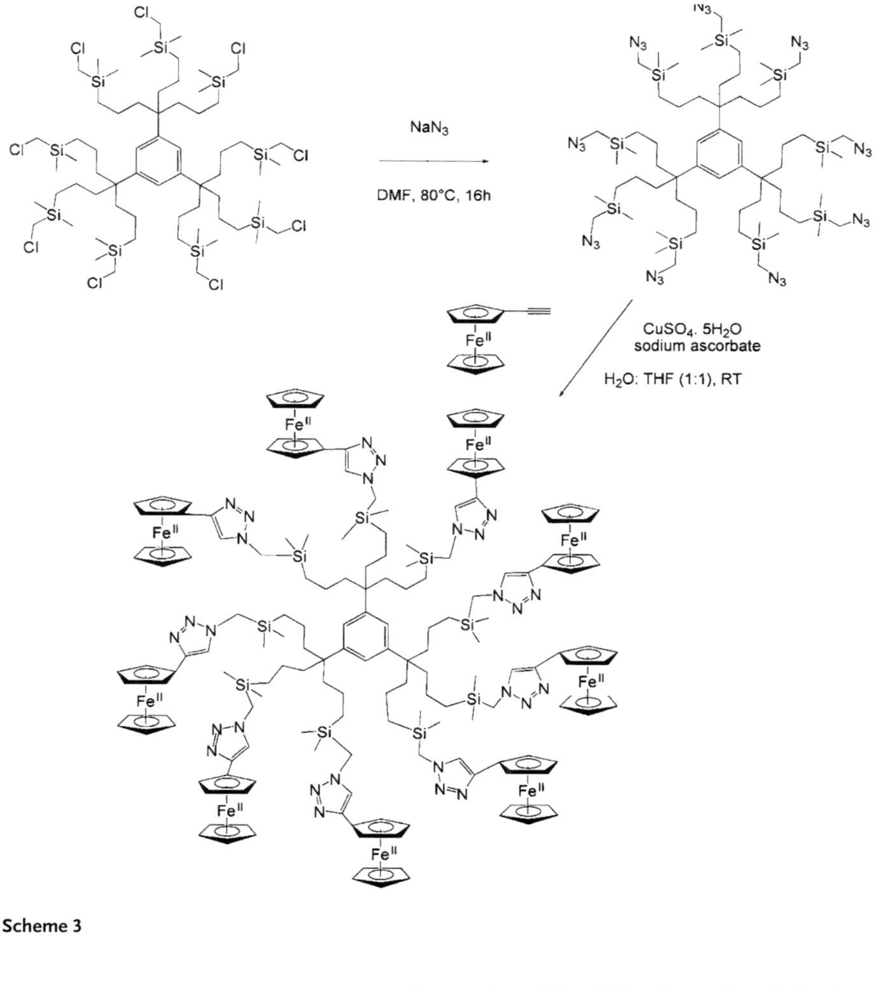

Scheme 3

Preparation of the 9-ferrocenyl-1,2,3-triazolyl dendrimer

Apparatus   Substitution of chloride by azide: one-necked Schlenk flask, reflux condenser, magnetic stirring bar. Click reaction: one-necked round-bottomed flask, magnetic stirring bar. Oil bath and hot magnetic plate. Safety glasses, laboratory coat, protective gloves.

Chemicals   Substitution of chloride by azide: 9-chloromethylsilyl-terminated dendrimer [9], sodium azide, DMF. Click reaction: THF, distilled water, dichloromethane, methanol, ethynylferrocene, copper sulfate pentahydrate, sodium ascorbate (1 mol L^{-1} aqueous solution), aqueous ammonia. Organic co-solvents need not be distilled for the »click« reaction. Complexes [FeCp(η^6-arene)][PF$_6$] are synthesized by reaction between ferrocene, the arene and AlCl$_3$ at 80–120 °C for 12 h according to procedures available and discussed in Ref. [7] of Chapter 40.

Attention!   Safety glasses, laboratory coat and protective gloves must be worn at all times in the laboratory.

Caution!   The reactions must be carried out in a well-ventilated hood. Dichloromethane and DMF are harmful. Do not breathe fumes and avoid contact with skin and eyes. Sodium azide and organic azides are explosive compounds when they are heated neat, so that extreme care must accompany their handling (no contact with metal spatula, no warm-up of neat azido derivative, operation exclusively under a well-protected hood, protective glasses not to be removed at any time when handling).

Procedure   **Synthesis of the 9-azidomethylsilyl dendrimer.** The 9-chloromethylsilyl-terminated dendrimer [9] (0.190 g, 0.130 mmol) and sodium azide (2 equiv. per branch) are heated at 80 °C for 16 h in dry DMF. The solvent is removed under vacuum, the crude product is dissolved in dichloromethane, washed twice with water, dried with sodium sulfate and filtered using paper, and the solvent is removed under vacuum. The dendrimer was precipitated using dichloromethane/methanol (1:100) and is obtained as a colorless waxy product in 99% yield (0.196 g, 0.129 mmol).

Characterization data: ^{1}H NMR (CDCl$_3$, 300 MHz): δ 6.94 (s, 3H, arom. CH), 2.72 (s, 18H, CH_2N$_3$), 1.62 (s, 18H, SiCH$_2$CH$_2$CH_2), 1.07 (s, 18H, SiCH$_2$CH_2CH$_2$), 0.58 (s, 18H, SiCH_2CH$_2$CH$_2$), 0.042 (Si(CH_3)$_2$) ppm. ^{13}C NMR (CDCl$_3$, 75.0 MHz): δ 145.7 (arom. C_q), 121.5 (arom. CH), 43.9 (SiCH$_2$CH$_2$CH$_2$), 41.9 (C_q), 41.1 (CH$_2$N$_3$), 17.7 (SiCH$_2$$CH_2CH_2$), 15.0 (SiCH$_2CH_2$$CH_2$), -4.04 (Si(CH$_3$)$_2$) ppm. ^{29}Si NMR (CDCl$_3$, 59.62 MHz): δ 3.33 (Si(CH$_3$)$_2$CH$_2$N$_3$) ppm. IR(cm^{-1}) : $\bar{\nu}$ 2093 (N$_3$). Calculated (found) for C$_{63}$H$_{129}$N$_{27}$Si$_9$: C, 49.86 (49.31); H, 8.57 (8.30)%.

General procedure for the »click« reactions: The azido dendrimer (1 equiv.) and the alkyne (1.5 equiv. per branch) are dissolved in tetrahydrofuran (THF) and water is added (1:1 THF/water). At 20 °C, CuSO$_4$ is added (1 equiv. per branch, 1 mol L^{-1} aqueous solution), followed by drop by drop addition of a freshly prepared solution of sodium ascorbate (2 equiv. per branch, 1 mol L^{-1} aqueous solution). The solution is allowed to stir for 30 min at room temperature. After removing THF under vacuum, dichloromethane and an aqueous ammonia solution are added. The mixture is allowed to stir for 10 min in order to remove all the Cu(I) trapped inside the dendrimer as [Cu(NH$_3$)$_6$]$^+$. The organic phase is washed twice with water, dried with sodium sulfate and filtered, and the solvent is removed under vacuum. The product is washed with pentane in order to remove the excess of alkyne and precipitated using dichloromethane/pentane (1:100).

Synthesis of the nonatriazolylferrocenyl dendrimer: The 9-ferrocenyltriazolyl dendrimer was synthesized from the 9-azidomethylsilyl dendrimer (0.100 g, 0.0659 mmol) and ethynylferrocene (0.187 g, 0.890 mmol) using the above general procedure for »click« reactions. The product was obtained as an orange powder in 94% yield (0.211 g, 0.0619 mmol).

Characterization data: ^{1}H NMR (CDCl$_3$, 300 MHz): δ 7.38 (s, 9H, triazole CH), 6.91 (s, 3H, Ar core), 4.70, 4.27 and 4.04 (s, 81H, Cp), 3.83 (s, 18H, SiCH_2N), 1.59 (s, 18H, CH_2CH$_2$CH$_2$Si), 1.06 (s, 18H, CH$_2$CH_2CH$_2$Si), 0.59 (s, 18H, CH$_2$CH$_2$CH_2Si), 0.045 (s, 54H, Si(CH_3)$_2$) ppm. ^{13}C NMR (CDCl$_3$, 75.0 MHz): δ 146.1 (C$_q$ of triazole), 120.0 (CH of triazole), 75.7 (C$_q$ of Cp), 69.5, 68.6 and 66.5 (CH of Cp), 43.8 (CH$_2$CH$_2$CH$_2$Si), 41.7 (benzylic C$_q$ of the core), 40.8 (CH$_2$N), 17.6 (CH$_2$$CH_2CH_2$Si), 14.8 (CH$_2CH_2$$CH_2$Si), -3.74 (Si(CH$_3$)$_2$) ppm. ^{29}Si NMR (CDCl$_3$, 59.6 MHz): δ 2.90 (Si(CH$_3$)$_2$CH$_2$N) ppm. MS (MALDI-TOF) m/z calcd. for C$_{171}$

$H_{219}Fe_9N_{27}Si_9$: 3 408.12, found: M^+ 3 408.18. Calculated (found) for $C_{171}H_{219}Fe_9N_{27}Si_9$: C, 60.26 (59.47); H, 6.48 (6.67)%. Polydispersity obtained by SEC < 1.02.

Waste Disposal DMF and THF are collected and disposed of in appropriately labeled containers. Aqueous layers containing sodium azide are collected and disposed of in an appropriately labeled container (sodium azide is harmful to the environment). These aqueous layers should never be mixed with acidic aqueous waste.

Explanation

The mechanism of the »click« reaction between terminal alkynes and azides starts with the formation of copper(I) acetylide (no reaction is usually observed with internal alkynes) followed by a stepwise annealing sequence via a six-membered copper-containing intermediate (see Scheme 4).

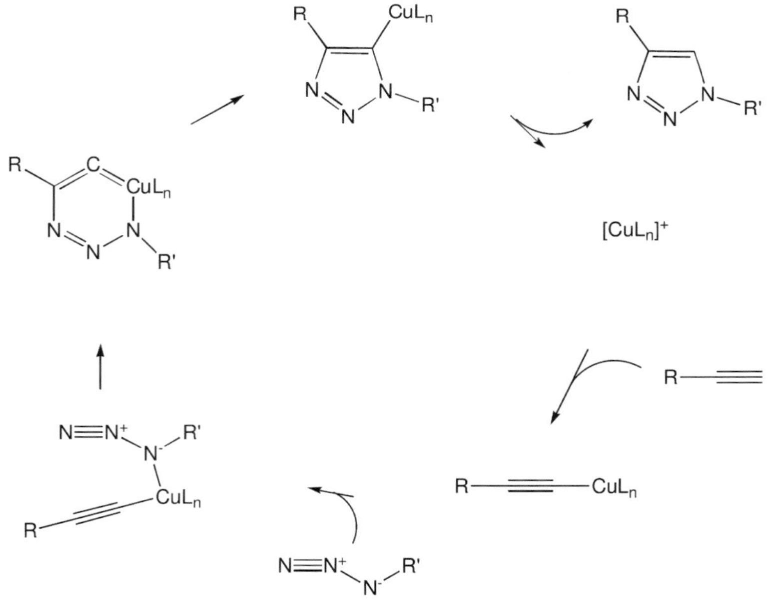

Scheme 4 Proposed mechanism for the Cu(I)-catalyzed cycloaddition of an azide and a terminal alkyne. The intermediates may involve a mono- copper species as drawn, or a ligand-bridged dicopper species.

The organocopper intermediates involved in the mechanism may be mono-copper ones or ligand-bridged dinuclear species. In the latter case, the interme-diate metallocycle would be a seven-membered ring instead of the six-mem-bered monocopper cycle drawn above [6]. When a tripodal monocopper catalyst is used, it is necessary that the catalytic copper species be monomeric [15], but when $CuSO_4$ and sodium ascorbate reductant are used, the lability of the Cu(I) species could occasionally favor ligand-bridged dinuclear intermediates [2, 6].

References

1 R. Huisgen, *Angew. Chem.* **1968**, *80*, 329; *Angew. Chem. Int. Ed. Engl.* **1968**, *7*, 321.
2 D. Astruc, *Organometallic Chemistry and Catal-ysis*, Springer, Berlin, **2007**, Chap. 21.
3 H.C. Kolb, M.G. Finn, K.B. Sharpless, *Angew. Chem.* **2001**, *113*, 2056; *Angew. Chem. Int. Ed.* **2001**, *40*, 2004.
4 (a) V.V. Rostovtsev, L.G. Green, V.V. Fokin, K.B. Sharpless, *Angew. Chem.* **2002**, *114*, 2708; *Angew. Chem. Int. Ed.* **2002**, *41*, 2596; (b) P. Wu, A.K. Feldman, A.K. Nugent, C.J. Hawker, A. Scheel, B. Voit, J. Pyun, J.M.J. Fréchet, K.B. Sharpless, V.V. Fokin, *Angew. Chem.* **2004**, *116*, 3951; *Angew. Chem. Int. Ed.* **2004**, *43*, 3928.
5 C.W. Tonøe, C. Christensen, M. Meldal *J. Org. Chem.* **2002**, *67*, 3057.
6 V.D. Bock, H. Hiemstra, J.H. van Maarse-veen, *Eur. J. Org. Chem.* **2006**, 51.
7 P.T. Anastas, J.C. Warner, *Green Chemistry: Theory and Practice*, Oxford University Press, New York, **1998**.
8 V. Sartor, L. Djakovitch, J.-L. Fillaut, F. Moulines, F. Neveu, V. Marvaud, J. Guittard, J.-C. Blais, D. Astruc, *J. Am. Chem. Soc.* **1999**, *121*, 2929.
9 J. Ruiz, G. Lafuente, S. Marcen, C. Ornelas, S. Lazare, E. Cloutet, J.-C. Blais, D. Astruc, *J. Am. Chem. Soc.* **2003**, *125*, 7250.
10 C. Ornelas, J.R. Aranzaes, E. Cloutet, S. Alves, D. Astruc, *Angew. Chem.* **2007**, *119*, 890; *Angew. Chem. Int. Ed.* **2007**, *46*, 872.
11 S. Badèche, J.-C. Daran, J. Ruiz, D. Astruc, *In-org. Chem.* **2008**, *47*, 4421.
12 (a) C. Ornelas, L. Salmon, J.R. Aranzaes, D. Astruc, *Chem. Commun.* **2007**, 4946; (b) C. Ornelas, J.R. Aranzaes, L. Salmon, D. Astruc, *Chem. Eur. J.* **2008**, *14*, 50.
13 A.K. Diallo, C. Ornelas, L. Salmon, J.R. Aran-zaes, D. Astruc, *Angew. Chem.* **2007**, *119*, 8798; *Angew. Chem. Int. Ed.* **2007**, *46*, 8644.
14 C. Ornelas, J. Ruiz, L. Salmon, D. Astruc, *Adv. Syn. Catal.* **2008**, *350*, 837.
15 N. Candelon, D. Lastécouères, A.K. Diallo, J.R. Aranzaes, D. Astruc, J.-M. Vincent, *Chem. Commun.* **2008**, 741.

28

Synthesis and Recycling of Six Different Nickel Complexes

Hartmut Schönberg and Hansjörg Grützmacher

Nickel complexes play an important role in chemical, technical and biological processes. The catalytic cleavage of dihydrogen at nickel metal surfaces was discovered in the 19th century. In homogeneous catalysis, nickel complexes play a predominant role as catalysts for the di- or oligomerization of olefins (nickel effect in the »Aufbaureaktion« discovered 1952 in Mülheim). Other applications of nickel complexes in processes of industrial importance are the carbonylation of ethylene to give propionic acid:

$$H_2C=CH_2 + CO + H_2O \rightarrow H_3C\text{-}CH_2COOH$$

or acrylic acid synthesis from acetylene, carbon monoxide and water:

$$HC \equiv CH + CO + H_2O \rightarrow H_2C=CHCOOH$$

Nickel complexes are also employed as catalysts for hydrocyanation and hydrosilylations of unsaturated hydrocarbons.

In addition, nickel complexes play very important roles in biological processes. In the enzyme urease, which is encountered in beans, the active center contains nickel and is responsible for the facile hydrolysis of urea to ammonia and carbon dioxide:

$$H_2N-CO-NH_2 + H_2O \rightarrow 2NH_3 + CO_2$$

Biological methane formation through micro-organisms – a critical process with respect to the greenhouse effect and the global warming – is mediated by the enzyme *methyl-coenzyme-reductase*, which contains nickel, probably in the oxidation state +1. A current working hypothesis is that a porphinoid Ni(I) complex catalyzes the overall reaction:

$$R'\text{-SH} + R''\text{S-CH}_3 \rightarrow R'\text{S-SR}'' + CH_4$$

About 10^8 tonnes of carbon monoxide per year are converted by microbes to carbon dioxide. Remarkably, these organisms use CO as the sole source of carbon and energy. Two of the three classes of carbon monoxide dehydrogenases contain nickel and catalyze the reaction, known as the water gas shift reaction in industrial processes.

$$CO + H_2O \rightarrow CO_2 + H_2$$

The activity of these enzymes is impressive and reaches up to 3200 molecules of CO per second – a rate that many man-made catalysts do not reach per hour!

Finally, hydrogenases – the basis of biological hydrogen technologies – also contain nickel as metal centers in the active site. These enzymes catalyze the oxidation of H_2 to protons, an important process in the energy production/consumption cycle in plants.

$$H_2 \rightarrow 2H^+ + 2e^-$$

A drawback in the application of nickel complexes is their intrinsically lower stability when compared to complexes made of metals from the fifth or sixth period. Any bond energy involving a transition metal from the fourth period is lower than the corresponding one involving a homologous metal center from the fifth or sixth period. This is one of the reasons why noble metal complexes with rhodium, iridium, palladium, platinum, and more recently also gold have become so popular in homogeneously catalyzed reactions. However, it would be of great economic and ecological benefit to have at hand either metal-free or nonnoble metal-mediated processes. Furthermore, in the not too distant future, catalytic reactions with noble metals may be confronted with serious resource problems. Therefore it is important to revitalize research activities which focus on the synthesis and application of fourth-row transition metal complexes, specifically taking into account the possible environmetal and ecological impact of these compounds.

The coordination chemistry of nickel(II) complexes is especially rich, and the coordination numbers 4, 5, and 6 are frequently encountered. All important structure types in coordination chemistry – tetrahedral, square planar, trigonal bipyramidal, square pyramidal, or octahedral – are observed, depending on the ligand field strength and/or steric requirements of the ligands. Octahedral, trigonal bipyramidal, square pyramidal and tetrahedral complexes are often green or blue and are paramagnetic (that is they have unpaired electrons). Square planar complexes are diamagnetic (that is all electrons are paired up) and have yellow to red colors. It is a speciality of nickel complexes that they do not necessarily follow the 18-electron rule, and electron-rich 20- and even 21-electron complexes are formed. Consequently, nickel complexes show a very rich redox chemistry with (formal) oxidation states at the nickel center ranging from +4 to −1.

The synthesis of nickel(II) complexes proceeds via consecutive ligand substitution reactions. The formation of various Ni(II) complexes can easily be followed by characteristic color changes. These equilibrium reactions and the stability of each complex are characterized by its formation constant. The experiments described here represent a chain of substitution reactions where a weak basic ligand is substituted by a stronger one, or a polydentate (chelate) ligand displaces a ligand which has fewer binding sites. In each case, the complex formation constant increases.

The experiment described below is based on the ideas developed by Fischer within the »Zurich model« for an ecological practical course in general chemistry [1, 2]. This model was considerably extended by Wiskamp [3]. Here, we report a slightly modified version which is performed in a so-called one-pot procedure with no loss of metal salt. It is the last step of the reaction sequence, the mixture being worked up in such a way that the nickel salt which was used as the starting material is recovered in more than 90% yield. Harmful waste is thus minimized.

Toxicology of Nickel: Nickel is one of the trace elements and is essential for plants. No nickel-dependent enzymes have yet been detected in mammals, but nickel is inevitably taken up and digested after the consumption of vegetable matter (indeed, vegetarians show higher nickel contamination). Up to 10% of the nickel content in food is taken up in the stomach and intestine (up to 25% if the organism lacks iron). In the blood plasma (Ni content < 1.2 µg L^{-1}), nickel is bonded and transported via proteins like albumin and amino acids like histidine. The total content in a 70 kg human amounts up to 15 mg. The metal and its inorganic compounds are considered to be comparatively nontoxic. However, some organometallic nickel compounds like the volatile $Ni(CO)_4$ are extremely toxic. Furthermore, nickel compounds may provoke allergenic reactions, especially contact allergies affecting the skin (dermatitis) and may also be carcinogenic/mutagenic. This is especially true for aerosols of nickel salts (e.g.

nickel acetate, nickel hydroxide, nickel sulfide, nickel oxide, nickel carbonate, etc.), and the maximum concentration should not exceed 0.05 mg Ni m^{-3}. The nickel-induced generation of oxy radicals is held responsible for its cancerogenic effect (DNA damage, blockage of DNA repair mechanisms). Intoxication with nickel leads to nausea, headache, breast pain, pulmonary edema, and brain hemorrhage. For these reasons, strict precautions must be taken when working with nickel compounds, and the release of large quantities into the environment must be avoided.

Preparation of Nickel Complexes

The individual steps 1–6 described below are illustrated in Figure 1. The formulae of the reagents used are displayed in Scheme 1.

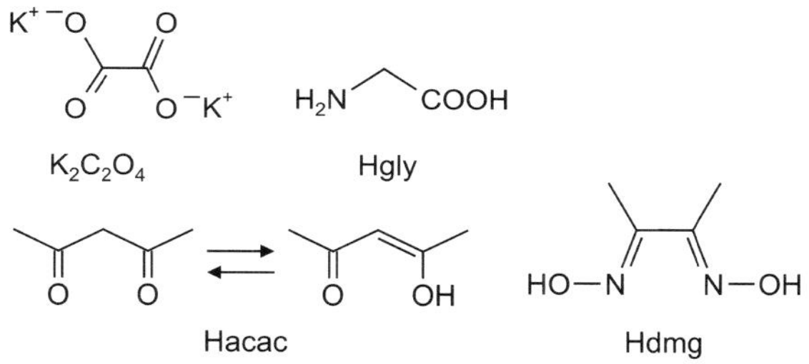

Scheme 1 Molecular formulae of the reagents used in the experiments: potassium oxalate ($K_2C_2O_4$); glycine (Hgly), acetylacetone (Hacac), dimethylglyoxime (Hdmg). Gly, acac, and dmg indicate the deprotonated forms.

Step 1: Preparation of [Ni(NH$_3$)$_6$]Cl$_2$

Apparatus   Two-necked 50 mL round-bottomed flask, magnetic stirrer equipped with a heating plate, oil bath, safety glasses, laboratory coat, protective gloves.

Chemicals   NiCl$_2$·6H$_2$O, distilled (demineralized) water, concentrated aqueous ammonia.

Attention!   Safety glasses and protective gloves must be worn at all times.

Caution!	Because of the allergenic property of nickel compounds and the strong skin irritating effect of ammonia, any contact with these chemicals must be avoided. All reactions must be carried out in a well-ventilated hood.

Procedure	$NiCl_2 \cdot 6H_2O$ (0.2 g, 0.84 mmol) is placed in a 50 mL round-bottomed flask and dissolved in 2 mL water. A concentrated aqueous solution of ammonia is added drop by drop. After a few drops have been added, a precipitation of nickel hydroxide, $Ni(OH)_2$, is observed, and the addition of ammonia is continued until this precipitate is completely dissolved and a clear deep blue solution of $[Ni(NH_3)_6]Cl_2$ has formed.

Step 2: Formation of $K_2[Ni(C_2O_4)_2(L)_2]$ (L = H_2O, NH_3)

Apparatus	Two-necked 50 mL round-bottomed flask, magnetic stirrer equipped with a heating plate, oil bath, reflux condenser, rubber septum, safety glasses, laboratory coat, protective gloves, reflux condenser, 5 mL plastic syringe, rubber septum, ice bath.

Chemicals	$K_2C_2O_4 \cdot H_2O$, distilled (demineralized) water.

Attention!	Safety glasses and protective gloves must be worn at all times.

Caution!	Because of the allergenic property of nickel compounds and the strong irritating effect of ammonia on the skin, any contact with these chemicals must be avoided. All reactions must be carried out in a well-ventilated hood.

Procedure	A reflux condenser is fitted to one neck of the flask containing the deep blue solution of $[Ni(NH_3)_6]Cl_2$, and the other neck is closed with the rubber septum. The reaction mixture is heated to 100 °C. Potassium oxalate, $K_2C_2O_4 \cdot H_2O$ (0.31 g, 1.7 mmol), is dissolved in a small amount of water (ca. 5 mL) and is then added by a syringe via the septum to the hot blue solution. Subsequently, the reaction mixture is cooled in an ice bath, whereupon turquoise crystals of a nickel oxalato complex, $K_2[Ni(C_2O_4)_2(L)_2]$, separate out (L = H_2O, NH_3).

Step 3: Formation of K[Ni(gly)$_3$]

Apparatus   Two-necked 50 mL round-bottomed flask, magnetic stirrer equipped with a heating plate, oil bath, reflux condenser, rubber septum, safety glasses, laboratory coat, protective gloves, 5 mL plastic syringe, Pasteur pipettes, heat gun.

Chemicals   Glycine (H$_2$NCH$_2$COOH), concentrated aqueous ammonia, distilled (demineralized) water.

Attention!   Safety glasses and protective gloves must be worn at all times.

Caution!   Because of the allergenic property of nickel compounds and the strong skin irritating effect of ammonia, any contact with these chemicals must be avoided. All reactions must be carried out in a well-ventilated hood.

Procedure   The reaction mixture obtained in Step 2 is warmed to room temperature and a solution of 0.185 g (2.46 mmol) glycine (H$_2$NCH$_2$COOH) in a small amount of water (ca. 5 mL) is added with a syringe. Subsequently, a few drops of a concentrated aqueous ammonia solution are added with a Pasteur pipette and the reaction mixture is briefly heated with a heat gun. A clear light blue solution of the amino acid nickel complex K[Ni(gly)$_3$] is formed.

Step 4: Formation of [Ni(acac)$_2$(H$_2$O)$_2$]

Apparatus   Two-necked 50 mL round-bottomed flask, magnetic stirrer equipped with a heating plate, oil bath, reflux condenser, rubber septum, safety glasses, laboratory coat, protective gloves, 2 mL plastic syringe.

Chemicals   Acetylacetone.

Attention!   Safety glasses and protective gloves must be worn at all times.

Caution!   Because of the allergenic property of nickel compounds and the strong skin irritating effect of ammonia, any contact with these chemicals must be avoided. All reactions must be carried out in a well-ventilated hood.

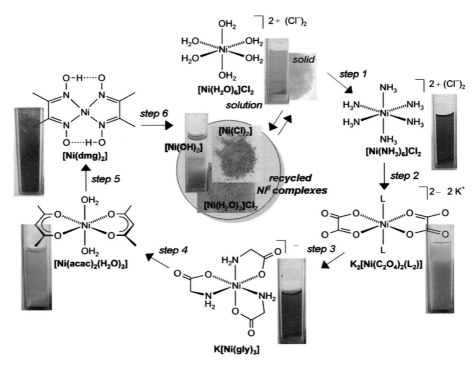

Figure 1 Illustration of the stepwise formation of nickel complexes in aqueous phase:

Step 1: $[Ni(H_2O)_6]Cl_2 + 6NH_3 \rightarrow [Ni(NH_3)_6]Cl_2 + 6H_2O$;

Step 2: $[Ni(NH_3)_6]Cl_2 + 2K_2C_2O_4 + 2L \rightarrow K_2[Ni(C_2O_4)_2(L)_2] + 6NH_3 + 2KCl$ (L = H_2O or NH_3);

Step 3: $K_2[Ni(C_2O_4)_2(L)_2] + 3Hgly + 3OH^- \rightarrow K[Ni(gly)_3] + 2C_2O_4{}^{2-} + K^+ + 3H_2O$;

Step 4: $K[Ni(gly)_3] + 2Hacac + 2H_2O \rightarrow [Ni(acac)_2(H_2O)_2] + Kgly + 2Hgly$;

Step 5: $[Ni(acac)_2(H_2O)_2] + 2Hdmg + 2OH^- \rightarrow Ni(dmg)_2 + 2acac^- + 4H_2O$

Step 6: Recycling by subsequent treatment with H_2SO_4, NaOH, and HCl.

Procedure

To the solution of $K[Ni(gly)_3]$, acetylacetone (0.236 mL, 2.4 mmol), acac, are added drop by drop using a 2 mL syringe. The reaction is stirred for 10 min, whereupon light blue crystals of nickel acetylacetonate, $[Ni(acac)_2(H_2O)_2]$, precipitate.

Step 5: Formation of Ni(dmg)₂

Apparatus

Two-necked 50 mL round-bottomed flask, magnetic stirrer equipped with a heating plate, oil bath, reflux condenser, rubber septum, safety glasses, laboratory coat, protective gloves, 2 mL plastic syringe.

| **Chemicals** | Diacetyldioxime, HON=C(Me)–C(Me)=NOH, dilute aqueous NaOH (10% by weight, ca. 2.5 M). |

| **Attention!** | Safety glasses and protective gloves must be worn at all times. |

| **Caution!** | Because of the allergenic property of nickel compounds and the strong skin irritating effect of ammonia, any contact with these chemicals must be avoided. All reactions must be carried out in a well-ventilated hood. |

| **Procedure** | To the suspension of $[Ni(acac)_2(H_2O)_2]$, a solution of diacetyldioxime (0.292 g, 2.52 mmol), HON=C(Me)–C(Me)=NOH, in 5 mL dilute aqueous sodium hydroxide (ca. 1.9 M) is added via a syringe. The light blue precipitate of $[Ni(acac)_2(H_2O)_2]$ disappears and a raspberry red precipitate of bis(diglyoximato)nickel(II), $Ni(dmg)_2$ is formed. This precipitate is highly insoluble and can be used to remove nickel(II) almost quantitatively from aqueous solutions at pH 5–12. |

Step 6: Workup and recycling of the nickel salt

| **Apparatus** | Two-necked 50 mL round-bottomed flask, magnetic strirrer equipped with a heating plate, oil bath, reflux condenser, rubber septum, safety glasses, laboratory coat, protective gloves, Pasteur pipettes, plastic syringe (5 mL), glass funnel and fluted filter papers, one-necked 50 mL round-bottomed flask, safety glasses, laboratory coat, protective gloves. |

| **Chemicals** | Concentrated sulfuric acid, powdered charcoal, H_2O_2 (30% by weight), aqueous NaOH (10% by weight), distilled (demineralized) water, concentrated HNO_3, (5% by weight) $BaCl_2$, concentrated aqueous HCl. |

| **Attention!** | Safety glasses and protective gloves must be worn at all times. |

| **Caution!** | Because of the allergenic property of nickel compounds and the strong skin irritating effect of ammonia, any contact with these chemicals must be avoided. All reactions must be carried out in a well-ventilated hood. |

Procedure   To the reaction mixture from Step 5, concentrated sulfuric acid is added carefully until the solution is strongly acidic (pH < 1). The solution shows a green color, indicating the formation of $[Ni(H_2O)_6]^{2+}$, and a precipitate may have formed. This precipitate consists of the organic ligand molecules ($H_2C_2O_4$, Hgly, Hacac, or Hdmg) and is removed by filtration using a glass funnel fitted with a fluted filter paper. Powdered charcoal (2 g) is added to the clear filtrate. The resulting mixture is heated to 100 °C and filtered while hot (glass funnel, fluted filter paper). The filtrate is then cooled to room temperature. Again, charcoal (1.5 g) is added and 30 wt% H_2O_2 (5 mL) is added via a syringe. This mixture is heated at 100 °C for about 1 h to remove and degrade traces of organic material. The charcoal is removed by filtration as described above and the filtrate is made basic (pH 11–12) by adding 10 wt% aqueous sodium hydroxide drop by drop with a Pasteur pipette. This causes the precipitation of nickel hydroxide, $Ni(OH)_2$, which is sparingly soluble at this pH and is collected by filtration (glass funnel with fluted filter paper). The compound must show a light green color (see Figure 1, middle) otherwise the separation of organic material was incomplete. The $Ni(OH)_2$ on the filter is rinsed several times with distilled (demineralized) water until no sulfate can be detected in the filtrate. (In order to test for this, a few drops of HNO_3 are added with a Pasteur pipette to make the water portion used in the washing process acidic and aqueous 5 wt% $BaCl_2$ is added. A white precipitate of $BaSO_4$ indicates that sulfate is still present, while no precipitation indicates that the wash water is sulfate free.) Subsequently, the $Ni(OH)_2$ is dissolved from the filter into a 50-mL round-bottomed flask using the minimum amount of concentrated HCl. The solution in the flask is placed in an oil bath at 150 °C and slowly evaporated to dryness. Higher temperatures must be avoided in order to prevent decomposition of the solid $NiCl_2$, which forms as a yellow powder (see Figure 1, middle). Exposure of this anhydrous $NiCl_2$ to the moisture in the air leads to the formation of green hexaaquonickel(II)dichloride $[Ni(H_2O)_6]Cl_2$, the starting material for the experiment. Careful working enables a yield of >95% (0.194 g, 0.81 mmol) to be obtained. Further purification can be carried out by recrystallization from H_2O.

References

1 H. Fischer, *Chemie in unserer Zeit* **1991**, *25*, 249.

2 P. Kating, H. Fischer, *Chemie in unserer Zeit* **1995**, *29*, 101.

3 V. Wiskamp, *Ausbildungsintegrierter Umweltschutz – vielseitige Weiterentwicklung des Zürcher Modells*, Fachhochschule Darmstadt, Fachbereich Chemie- und Biotechnologie, **1997**.

Part IV:
Limiting Waste and Exposure

29

Disposal of Sodium and Potassium Residues [1]

Herbert W. Roesky

Apparatus   Ceramic flower pot (10 cm in diameter), big porcelain dish, pair of tweezers, safety glasses, laboratory coat, protective gloves

Chemicals   Water, sodium and potassium residues.

Attention!   Alkali metals and water react intensely, producing fire. Safety glasses and protective gloves must be worn at all times.

Disposing of sodium and potassium residues often causes accidents due to the peroxides formed on the surface of alkali metals when they are kept in the air for some time.

Sodium and potassium are often used for drying ether, tertiary amines, hydrocarbons and aromatics. Sodium and potassium residues have to be disposed of. Usually this is done with 2-propanol. Serious accidents can occur if peroxide-containing alkali metals are pressed or cut, and if an alcohol containing water or an alcohol of low molar mass is used. Often the reaction cannot be controlled when the alcohol is added too fast.

Procedure   The following method describes a safe procedure that is less expensive and environmentally friendly.

The bottom of a flower pot is covered (inside) with a paper filter, and it is half-filled with sand. About 1 g of an alkali metal residue is placed on top of the sand using a pair of tweezers. Then the pot is completely filled with sand. The pot is then placed in the dish, and water is added to a depth of 2 cm. Because of the capillary action of the sand, the water rises in the pot, and after some time the sand surface becomes wet and darkens in color. After one or two days the water will have decomposed the alkali metals. The sand is washed and dried, and can be used again. This method avoids the use of a flammable alcohol.

Waste Disposal   The water used for washing the sand can be flushed down the drain.

Reference

1 H.W. Roesky, *Inorg. Chem.* **2001**, *40*, 6855.

30

Juglone (5-Hydroxy-1,4-naphthoquinone)

Kieran Joyce, Emma E. Coyle, Jochen Mattay and Michael Oelgemöller

It has been known for hundreds of years that herbal preparations of the leaves, bark, wood and husks of the walnut plant provide relief from bacterial and fungal diseases of the skin. However, the therapeutic agent, juglone, was not isolated until 1856 by Vogel and Reischauer [1]. Once it was isolated, studies began on juglone and its derivatives and a large number of therapeutic affects were discovered [2]. Today we know a lot about juglone, including its allelopathic activity and its cytotoxicity to plant, microbial, aquatic and mammalian life. Because of these properties, juglone is a versatile building block for the synthesis of biologically active compounds such as *Urdamycinone B* [3] and *Frenolicin B* [4]. However, a sufficient and especially a sustainable method for its synthesis has not yet been established. Existing methods include the chemical oxidation of 1,5-dihydroxynaphthalene. In particular, current methods use inorganic oxidants such as active manganese dioxide (MnO_2) [5], sodium dichromate ($Na_2Cr_2O_7$) [6], and thallium trinitrate (TTN) [7]. Periodic acid (HIO_4) has also been used [8]. In most of these cases the yield has been low with the exception of that using TTN (yield ~64%). However, despite the relatively low yields another more serious problem still remains. The reagents required are themselves highly toxic to the environment and the side products produced can also be hazardous. The following procedure has been researched in an effort to produce juglone via an environmentally friendly process. Juglone can be synthesized from 1,5-dihydroxynaphthalene in good yield without the use or production of environmentally toxic chemicals if visible light, air and a common sensitizer (Rose Bengal) are used Scheme 1.

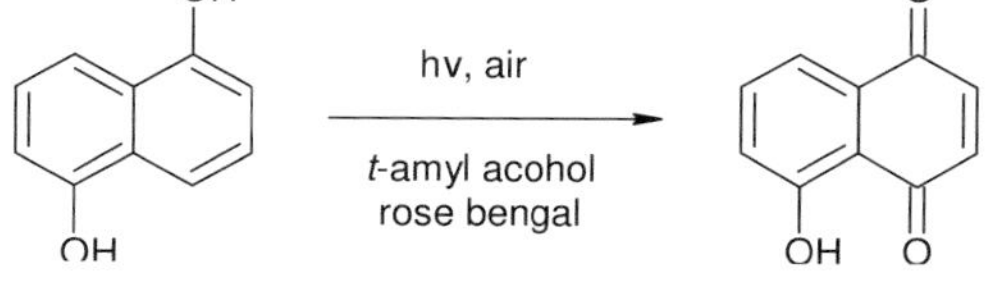

Scheme 1

Preparation of juglone

Apparatus 100 mL Schlenk flask, cold finger, magnetic stirrer, air pump/supply, 3 mm diameter chemical resistant tubing, HPLC inlet filter, 500 W halogen lamp, silica plates, safety glasses, laboratory coat, protective gloves.

Chemicals 1,5-dihydroxynaphthalene, Rose Bengal (4,5,6,7-tetrachloro-2',4',5',7'-tetraiodofluorescein), 2-methyl-2-butanol, cyclohexane, ethyl acetate.

Attention! Safety glasses and protective gloves must be worn at all times. Eyes should not be exposed to bright light.

Caution! The sensitizer Rose Bengal can easily stain the skin upon contact. It can also cause photodynamic therapy effects on the skin upon exposure to light. Juglone itself has allelopathic properties. Release to the environment could cause both aquatic and environmental problems [9, 10].

Procedure 1,5-Dihydroxynaphthalene (0.16 g, 1 mmol) is dissolved in 2-methyl-2-butanol (100 mL) and placed in a 100 mL Pyrex Schlenk flask. Rose Bengal (50 mg, 0.05 mmol) is also added, and the Schlenck flask is placed in front of a 500 W halogen lamp. The solution is cooled via a cold finger and running water. Air is then bubbled through the solution to supply oxygen for the reaction. This is done using 3 mm diameter tubing and an HPLC inlet filter to generate fine bubbles (Figures 1 and 2). The reaction is then initiated by turning on the halogen lamp. The reaction is followed via thin-layer chromatography (TLC) analysis using silica plates and a 3:1 mixture of cyclohexane/ethyl acetate as the mobile phase. Juglone shows a bright orange spot with an R_f value of 0.52. ^{1}H NMR analysis can be used instead, but is more time consuming. The reaction is deemed complete after 2 h.

Workup: The solvent (2-methyl-2-butanol) is removed by rotary evaporation (30 mbar and 40 °C). When almost all of the solvent is removed, 1–2 spatula tips of silica gel are added to the crude product solution. The remaining solvent is removed by rotary evaporation. The silica is added to aid the removal of the crude product from the

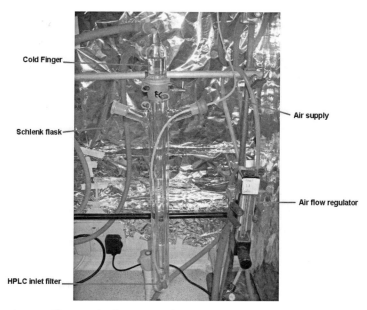

Cold Finger

Air supply

Schlenk flask

Air flow regulator

HPLC inlet filter

Figure 1 Photograph of experimental set-up.

round-bottomed flask. The crude product is then purified by column chromatography using silica gel as the solid phase. The mobile phase is a 3:1 mixture of cyclohexane/ethyl acetate. The eluted fractions are compared via TLC analysis and similar fractions are combined. R_f(product) = 0.52. The solvent is removed by rotary evaporation to yield 108.8 mg (0.62 mmol, 62% yield) of bright orange crystals.

Characterization data: m.p. 158–160 °C. ^{1}H NMR (400 MHz, CDCl$_3$): δ 6.96 (s, 2H, H$_{quinone}$), 7.29 (dd, ^3J = 7.5 Hz, ^4J = 2.2 Hz, 1H, H$_{arom}$) 7.62 (dd, ^3J = 7.5 Hz, ^4J = 2.2 Hz, 1H, H$_{arom}$), 7.65 (dd, ^3J = 7.5 Hz, 1H, H$_{arom}$), 11.92 (s, 1H, -OH) ppm. ^{13}C NMR (100 MHz, CDCl$_3$): δ 114.0, 118.1, 123.5, 130.8, 135.5, 137.6, 138.6, 160.6, 183.2, 189.3 ppm. IR (KBr): $\bar{v}$ 3400, 3058, 1662, 1641, 1590, 1448, 1289, 1225, 1151, 1098, 1081, 863, 827, 762, 703 cm^{-1}. MS (70eV): m/z 174(M$^+$, 100%). Calculated (found) for C$_{10}$H$_6$O$_3$: C, 68.98 (68.25); H, 3.47 (3.70)%.

Waste Disposal   The 2-methyl-2-butanol used as solvent for the reaction can be recycled. Once the solvent is removed from the product by rotary evaporation under vacuum it can be collected and decolorized by stirring

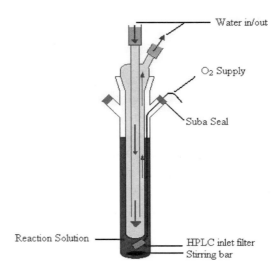

Water in/out

O₂ Supply

Suba Seal

Reaction Solution — HPLC inlet filter
Stirring bar

Figure 2 Diagram of experimental set-up.

with 2–3 spatula tips of activated charcoal. This can be left overnight. The solvent can then be filtered to remove the charcoal and is ready for use again. ¹H NMR can be used to verify solvent purity.

The same procedure can be carried out for the mobile phase. However, its use as a mobile phase in column chromatography is limited to crude compounds that require a 3:1 cyclohexane/ethyl acetate mixture for good separation of fractions.

The solid waste generated consists mainly of silica gel, Rose Bengal and small quantities of unreacted starting material. This solid waste once dry can be placed in the laboratory's solid waste container. The laboratory should hire a licensed contractor to dispose of the waste.

Explanation

Current methods for the synthesis of juglone involve or chemical oxidation of 1,5-dihydroxynaphthalene. Examples of such chemical oxidants are active manganese dioxide (MnO_2) [5], sodium dichromate ($Na_2Cr_2O_7$) [6] and thallium trinitrate (TTN) [7]. Once released to the environment, these chemicals can have serious toxic affects. However, the described method for production of juglone does not require the use of any such chemicals, which leads to very little waste production during the chemical process. Rose Bengal, a common dye with no known toxic effects is the only waste chemical generated. It is usual to generate singlet oxygen as the actual oxidant. However, it may be immobilized on a solid support for recovery and reuse [11]. All solvents can be recycled and reused in further juglone synthesis reactions. The solvent 2-methyl-2-butanol was used as an alternative to the commonly used CH_2Cl_2/methanol mixture [12]. Again, this is far more environmentally friendly than the chlorinated solvent CH_2Cl_2.

It is important to note that the juglone was synthesized using artificial light from a halogen lamp. The reaction can also be carried out using natural light. If direct sunlight is available, the reaction set-up can be placed out in the sun, and the reaction proceeds as normal. In fact, both the reaction time and the yield improve when using natural light. If solar concentrators are used, the reaction time can be drastically reduced. Example of such solar reactors can be seen at the German Aerospace Center (DLR) in Cologne in Germany [13, 14]. The use of these solar reactors has proved that the production of certain fine chemicals can be carried out on an industrial scale using no more than sunlight, air and nontoxic solvents and sensitizers.

References

1 R.H. Thomson, *J. Org. Chem.* **1948**, *13*, 377 (and references therein).
2 R.H. Thomson, *Naturally Occuring Quinones*, Academic Press, London and New York, **1971**.
3 G. Matsuo, Y. Miki, M. Nakata, S. Matsumura, K. Toshima, *J. Org. Chem.* **1999**, *64*, 7101.
4 P. Contant, M. Haess, J. Riegl, M. Scalone, M. Visnick, *Synthesis* **1999**, 821.
5 L.K. Kumari, M. Pardhasaradhi, *Ind. J. Chem.* **1982**, *21*(B), 1067.
6 R.G. Jesaitis, A. Krantz, *J. Chem. Educ.* **1972**, *49*, 436.
7 D.J. Crouse, M.M. Wheeler, M. Goemann, P.S. Tobin, S.K. Basu, D.M.S. Wheeler, *J. Org. Chem.* **1981**, *46*, 1814.

8 A.V. Pinto, V.F. Ferreira, M. do Carmo, F.R. Pinto, *Synth. Comm.* **1985**, *15*, 1177.
9 J. Shibu, A.R. Gillespie, *Plant Soil* **1998**, *203*, 199.
10 A.M. Hejl, K.L. Koster, *J. Chem. Ecol.* **2004**, *30*, 453.
11 O. Suchard, R. Kane, B.J. Roe, E. Zimmermann, C. Jung, P.A. Waske, J. Mattay, M. Oelgemöller, *Tetrahedron* **2006**, *62*, 1467.
12 J. Griffiths, K.-Y. Chu, C. Hawkins, *J.C.S. Chem. Comm.* **1976**, 676.
13 M. Oelgemöller, N. Healy, L. de Oliveira, C. Jung, J. Mattay, *Green Chem.* **2006**, *8*, 831.
14 M. Oelgemöller, C. Jung, J. Mattay, *Pure. Appl. Chem.* **2007**, *79*, 1939.

31
Fluor Retard

Herbert W. Roesky

In 1529 Agricola described the use of fluorspar as a flux. This mineral was important for the preparation of metals. About 300 years later Ampère wrote to Davy suggesting the name *le fluore* (fluorine) for the presumed new element in CaF_2 and hydrogen fluoride. The term fluorescence was coined by Stokes for light emission from fluorspar when it was heated. However, all attempts to isolate the element fluorine failed because of its extreme reactivity. In 1886 Moissan electrolyzed a cooled solution of KHF_2 in anhydrous liquid HF and reported his results to the Academy: »One can indeed make various hypotheses on the nature of the liberated gas; the simplest would be that we are in the presence of fluorine, but it would be possible, of course, that it might be a perfluoride of hydrogen or even a mixture of hydrofluoric acid and ozone ...« Since that time, elemental fluorine, hydrogen fluoride, and many other fluorinating reagents have been prepared for the synthesis of new fluorine-containing compounds. We have discovered that Me_3SnF and Ph_3SnF are very efficient reagents for metathetical reactions with organometallic compounds. Me_3SnF, known as Fluor Retard, is used for nonvolatile compounds, whereas Ph_3SnF is used for volatile fluorides.

Two preparations using Me₃SnF

Preparation of (η^5-C₅Me₅)TiF₃

Apparatus	100 mL Schlenk flask with septum, magnetic stirrer, T-shaped N_2 inlet, syringe (50 mL), safety glasses, laboratory coat, protective gloves.
Chemicals	Me_3SnF, (η^5-C₅Me₅)TiCl₃, dry toluene.

Attention!	Safety glasses and protective gloves must be worn at all times.

Caution!	Because of their toxicity and volatility, care should be taken to avoid inhalation of toluene, trimethyltin chloride or trimethyltin fluoride, or contact of their solutions with the skin. All reactions should be carried out in a well-ventilated hood.

Procedure	An oven-dried two-necked 100 mL Schlenk flask equipped with a magnetic stirrer, a septum, and a T-shaped N_2 inlet is charged in a dry box with freshly sublimed Me_3SnF (2.75 g, 15 mmol) and $(\eta^5-C_5Me_5)TiCl_3$ (1.45 g, 5 mmol). Toluene (50 mL) is added and the resulting mixture is stirred at room temperature for 4 h. During this time the suspension disappears and a clear solution is formed. The disappearance of the Me_3SnF suspension is an indicator of the progress of the reaction. The solvent and the Me_3SnCl are removed in vacuum. The orange residue is sublimed at 110 °C / 10^{-2} mbar to yield 1.1 g (92 %) of orange $(\eta^5-C_5Me_5)TiF_3$.

Characterization data: m.p. 180 °C. ^1H NMR: δ 1.93 (s, C_5Me_5) ppm. ^{19}F NMR: δ 124.0 (s) ppm. IR (CsI, cm^{-1}): $\bar{v}$ 1072, 1023, 704, 648, 595, 590, 486, 341 MS: m/z 240 (M$^+$, 38 %), 135 (C_5Me_5, 100 %).

Preparation of [LCaF(thf)]$_2$ L = CH[CMe(2,6-*i* Pr$_2$C$_6$H$_3$N)]$_2$

Apparatus	100 mL Schlenk flask with septum, magnetic stirrer, syringe (50 mL), T-shaped N_2 inlet, safety glasses, laboratory coat, protective gloves.

Chemicals	[LCaN(SiMe$_3$)$_2$(thf)], Me$_3$SnF, dry THF, dry hexane.

Attention!	Safety glasses and protective gloves must be worn at all times.

Procedure	The Schlenk flask is charged in a dry box with [LCaN(SiMe$_3$)$_2$(thf)] (1.38 g, 2.0 mmol) and Me$_3$SnF (0.368 g, 2.0 mmol). After adding THF (40 mL), the mixture is stirred for 16 h. All the volatiles are removed in vacuum and the residue is extracted with hexane (60 mL). The solution upon concentration in vacuum afforded [LCaF(thf)]$_2$ as a white solid. The solid can be recrystallized from a mixture of hot toluene and THF. Yield 0.65 g (60 %).

Characterization data: m.p. 277–280 °C. ^{1}H NMR (500 MHz, C_6H_6): δ 7.09 (s, 12 H, *m*-, *p*-Ar-*H*), 4.72 (s, 2H, γ-C*H*), 3.54 (m, 8H, O–C*H*$_2$–CH$_2$) 3.14 (sept, 8H, C*H*(Me)$_2$), 1.64 (s, 12H, C*H*$_3$), 1.42 (m, 8H, O-CH$_2$-C*H*$_2$), 1.24–1.22 (d, 24H, CH(C*H*$_3$)$_2$), 1.02-1.01 (d, 24H, CH(C*H*$_3$)$_2$) ppm. ^{19}F NMR (188.77 MHz, C_6D_6): δ –78 ppm. MS (70eV): m/z 202 (100 %, 2,6-*i* Pr$_2$C$_6$H$_3$NCMe).

Explanation

Fluor Retard is easily available using Me_3SnCl and NaF in water. It is quantitatively obtained, and the resulting Me_3SnCl after the fluorination reaction can easily be removed in vacuum and reconverted with aqueous NaF to Me_3SnF. This is a striking improvement over other fluorinating reagents. In addition, this elegant method of fluorination uses stoichiometric amounts and can be applied to complexes containing M–C, M–O, and M–N bonds (M = metal). Even Ti(III) compounds can be converted to the corresponding fluorides without changing the oxidation state of the metal.

The name Fluor Retard characterizes the process of the fluorination that is due to the low solubility of Me_3SnF in organic solvents and its conversion to Me_3SnCl. $(\eta^5$-$C_5Me_5)TiF_3$ is a highly active catalyst producing syndiotactic polystyrene with a very high melting point (277 °C) and high molecular mass (660 000 g mol^{-1} at 50 °C). It shows thermal stability up to the polymerization temperature of 70 °C. The chlorinated analog has strongly reduced activity and results in a polymer of lower molecular weight.

As early as 1529 Agricola reported on the use of CaF_2 as a flux. CaF_2 is the raw material for the preparation of nearly all fluorine compounds. It reacts with sulfuric acid with elimination of HF. The reported [LCaF(thf)]$_2$ is the only derivative of CaF_2 so far known. It can be used for the preparation of thin layers of CaF_2 on a support at room temperature. CaF_2 is transparent to light in the UV and visible region.

References

1 A. Herzog, F.-Q. Liu, H.W. Roesky, A. Demsar, K. Keller, M. Noltemeyer, F. Pauer, *Organometallics* **1994**, *13*, 1251.

2 H.W. Roesky, A. Herzog, F.-Q. Liu, *J. Fluorine Chem.* **1995**, *71*, 161.

3 H.W. Roesky, A. Herzog, F.-Q. Liu, *J. Fluorine Chem.* **1995**, *72*, 183.

4 F.-Q. Liu, A. Herzog, H.W. Roesky, I. Usón, *Inorg. Chem.* **1996**, *35*, 741.

5 E.F. Murphy, T. Lübben, A. Herzog, H.W. Roesky, A. Demsar, M. Noltemeyer, H.-G. Schmidt, *Inorg. Chem.* **1996**, *35*, 23.

6 E.F. Murphy, P. Yu, S. Dietrich, H.W. Roesky, E. Parisini, M. Noltemeyer, *J. Chem. Soc. Dalton Trans.* **1996**, 1983.

7 S.A.A. Shah, H. Dorn, A. Voigt, H.W. Roesky, E. Parisini, H.-G. Schmidt, M. Noltemeyer, *Organometallics* **1996**, *15*, 3176.

8 M.G. Walawalkar, R. Murugavel, H.W. Roesky, *Eur. J. Solid State Inorg. Chem.* **1996**, *33*, 943.

9 (a) E. Krause, *Ber. Dtsch. Chem. Ges.* **1918**, *51*, 1447; (b) W.K. Johnson, *J. Org. Chem.* **1960**, *25*, 2253; L.E. Levchuk, J.R. Sams, F. Aubke, *Inorg. Chem.* **1972**, *11*, 43.

10 A.M. Cardoso, R.J.H. Clark, S. Moorhouse, *J. Chem. Soc. Dalton Trans.* **1980**, 1156.

11 H.W. Roesky, A. Herzog, K. Keller, *Z. Naturforsch.* **1994**, *49b*, 981.

12 W. Kaminsky, S. Lenk, V. Scholz, H.W. Roesky, A. Herzog, *Macromolecules* **1997**, *30*, 7647.

13 S. Nembenna, H.W. Roesky, S. Nagendran, A. Hofmeister, J. Magull, P.-J. Wilbrandt, M. Hahn, *Angew. Chem.* **2007**, *119*, 2564; *Angew. Chem. Int. Ed.* **2007**, *46*, 2512.

32

Dicopper(I) Oxalate Complexes as Molecular Precursors for Copper Deposition

Michael Stollenz and Franc Meyer

Copper deposition has become an important and rapidly growing area in the manufacture of integrated circuits, since this metal is the first choice as a conducting material in the microelectronics industry. Favorable properties of copper include its reliability in carrying high-current densities and its ease of handling compared to alternatives like aluminum or tungsten. Well-defined molecular copper(I) and copper(II) complexes are thus required as precursors for pure, thin copper films prepared by various deposition techniques such as chemical vapor deposition (CVD), aerosol-assisted CVD, atomic layer deposition, or spin coating. In the case of copper(II) complexes, however, an additional reducing agent is usually required, and the conversion to elemental copper is often insufficient. Copper(I) precursors of the general type LCu(I)(β-diketonate), where the β-diketonate is usually hexafluoroacetylacetonate and L a Lewis base such as a phosphine, alkyne or alkene, have received much attention, because these complexes decompose in a clean disproportionation to afford $Cu(II)(\beta\text{-diketonate})_2$, the free ligand L and metallic copper according to

$$2\,LCu^I(\beta\text{-diketonate}) \xrightarrow{\Delta T} Cu^0 + Cu^{II}(\beta\text{-diketonate})_2 + 2\,L$$

Major drawbacks in this process are the limited yield of elemental copper (up to a maximum of 50%) and the poor adhesion of the copper films onto the TiN diffusion barrier layer when using fluorine-containing precursors. Searching for new self-reducing and fluorine-free precursors, we discovered and investigated some copper(I) oxalate complexes with neutral Lewis bases as capping ligands. These molecular compounds are easily prepared, reasonably air stable, and highly soluble. They incorporate the oxalate ligand as a built-in reducing agent that ensures a 100% yield of clean metallic copper upon thermal decomposition via an internal redox process, with the intact and recyclable free Lewis bases and nontoxic CO_2 as the only byproducts:

$$\text{L—Cu} \overset{O \cdots O}{\underset{O \cdots O}{\diagdown}} \text{Cu—L} \quad \xrightarrow{\triangle T} \quad 2\,Cu^0 + 2\,L + 2\,CO_2$$

Two preparations of dicopper(I) oxalate complexes

Preparation of bis{[bis(trimethylsilyl)acetylene]copper(I)} oxalate (1)

Apparatus
100 mL Schlenk flask with septum, magnetic stirrer, syringe (50 mL), safety glasses, laboratory coat, protective gloves.

Chemicals
Cu_2O, $Me_3SiC\equiv CSiMe_3$, oxalic acid, dry dichloromethane.

Attention!
Safety glasses must be worn at all times. Protective gloves are recommended.

Caution!
Because of their harmfulness and/or volatility, care should be taken to avoid inhalation or contact of dichloromethane, copper(I) oxide, and oxalic acid with the skin. Bis(trimethylsilyl)acetylene and 3-hexyne are irritant and flammable. All reactions should be carried out in a well-ventilated hood.

Procedure
A 100 mL Schlenk flask with a magnetic stirrer is loaded with Cu_2O (3.4 g, 23.8 mmol), oxalic acid (2.1 g, 23.3 mmol), and $Me_3SiC\equiv CSiMe_3$ (8.0 g, 46.9 mmol), purged with argon and sealed with a septum. Treatment with dichloromethane (60 mL) results in a red suspension, which is stirred at room temperature for 4 h while the solution becomes almost colorless. Subsequently, the remaining Cu_2O is filtered off and the resulting clear solution is evaporated to dryness (0.01 mbar) to leave a colorless solid. Recrystallization from dichloromethane at –20 °C yields colorless crystals of **1** (11.7 g, 90%).

Characterization data: m.p. 153 °C (decomposition). ^1H NMR (CDCl$_3$): δ 0.30 (s, SiMe$_3$) ppm. $^{13}C\{^1H\}$ NMR: δ 0.0 (SiMe$_3$), 114.2 (C$\equiv$C), 171.8 (COO) ppm. IR (KBr, cm^{-1}): $\bar{\nu}$ 1935 (m, C$\equiv$C), 1642 (s, COO), 1354 (w), 1309 (m). MS: m/z 788 ([M + Cu(Me$_3$SiC$\equiv$CSiMe$_3$]$^+$, 25 %), 618 ([M + Cu]$^+$, 10%), 403 ([M – CuO$_4$C$_2$]$^+$, 68%), 233 ([M – (Me$_3$SiC$\equiv$CSiMe$_3$)CuO$_4$C$_2$]$^+$, 100%). Calculated (found) for $C_{18}H_{36}Cu_2O_4Si_4$: C, 38.9 (38.7); H, 6.5 (6.6)%.

Preparation of bis[(3-hexyne)copper(I)] oxalate (2)

Apparatus	100 mL Schlenk flask with septum, magnetic stirrer, syringe (50 mL), safety glasses, laboratory coat, protective gloves.
Attention!	Safety glasses must be worn at all times. Protective gloves are recommended.
Chemicals	Cu_2O, EtC≡CEt, oxalic acid, dry dichloromethane.
Procedure	Analogously to the preparation of **1**, Cu_2O (1.8 g, 12.6 mmol), oxalic acid (1.1 g, 12.2 mmol), EtC≡CEt (3 mL, 26.4 mmol), and dichloromethane (50 mL) are used. Recrystallization from dichloromethane at −20 °C yields colorless needles of **2** (2.6 g, 56%). *Characterization data:* ^1H NMR ($CDCl_3$): δ 1.22 (t, 12H, $^3J = 7.3$ Hz, CH_3), 2.49 (q, 8H, $^3J = 7.3$ Hz, CH_2) ppm. $^{13}C\{^1H\}$ NMR: δ 14.4 (CH_3), 15.5 (CH_2), 87.6 (C≡C), 171.4 (COO) ppm. IR (KBr, cm^{-1}): $\bar{v}$ 2053, 2020 (w, C≡C), 1645 (vs, COO), 1355 (w), 1314 (m). MS: m/z 525 ([M + Cu(EtC≡CEt)]⁺, 7%), 443 ([M + Cu]⁺, 7%), 227 ([M − CuO_4C_2]⁺, 100%), 145 ([M − (EtC≡CEt)CuO_4C_2]⁺, 96%). Calculated (found) for $C_{14}H_{20}Cu_2O_4$: C, 44.3 (44.7); H, 5.3 (4.9)%.
Waste Disposal	The excess copper(I) oxide separated by filtration is dried and disposed of in the collecting box for metal-containing waste. Dichloromethane is first dried with CaH_2 in order to remove the content of water originating from the reaction. For reuse, the solvent is then redistilled under inert conditions, and its purity is finally checked by gas chromatography.

Explanation

In the presence of potential Lewis bases such as olefins or alkynes, copper(I) oxide and oxalic acid react with concomitant elimination of water to give well-defined binuclear copper(I) oxalate complexes. The thermal decomposition behavior of **1** and **2** demonstrates their high potential as precursors for copper deposition. On the one hand, both complexes are easy to synthesize and also easy to handle, since they are stable up to 60 °C and 100 °C, respectively. Furthermore, they exhibit rather low volatility. On the other hand, beyond this temperature range **1** and **2** start to decompose in well-behaved internal redox processes that result in the formation of elemental copper and CO_2, with an initial loss of both neutral alkyne ligands. The yield of copper is quantitatively related to the precursor complexes, the only clean by products are a nontoxic gas, and the volatile co-ligands, which can be easily recycled. Overall, these dicopper(I) oxalate complexes are excellent candidates for environmentally friendly copper deposition applications.

References

1 K. Köhler, J. Eichhorn, F. Meyer, D. Vidovic, *Organometallics* **2003**, *22*, 4426.
2 K. Köhler, F. Meyer, (Merck Patent GmbH) WO 2004/000850.
3 J. Teichgräber, S. Dechert, F. Meyer, *J. Organomet. Chem.* **2005**, *690*, 5255.
4 P. Doppelt, *Coord. Chem. Rev.* **1998**, *178-180*, 1785.
5 J. Rickerby, J.H.G. Steinke, *Chem. Rev.* **2002**, *102*, 1525.
6 T.-Y. Chen, J. Vaissermann, E. Ruiz, J. P. Senateur, P. Doppelt, *Chem. Mater.* **2001**, *13*, 3993.
7 A. Jain, K.-M. Chi, T.T. Kodas, M.J. Hampden-Smith, J.D. Farr, M.F. Paffett, *Chem. Mater.* **1991**, *3*, 995.
8 D. Hnyk, M. Bühl, P.T. Brain, H.E. Robertson, D.W.H. Rankin, *J. Am. Chem. Soc.* **2002**, *124*, 8078.
9 Y. S. Kim, Y. Shimogaki, *J. Vac. Sci Technol. A* **2001**, *19*, 2642.
10 P.-F. Hsu, Y. Chin, T.-W. Lin, C.-S. Liu, A.J. Carty, S.-M. Peng, *Chem. Vap. Deposition* **2001**, *7*, 28.

33

Environmentally Friendly Recycling of Sodium [1]

Herbert W. Roesky

In 1807 Sir Humphry Davy isolated sodium by passing an electric current through molten sodium hydroxide.

Sodium is a silvery white metal that quickly oxidizes in air and reacts vigorously with water. It melts at 97.8 °C and is lighter than water. Sodium is present in great quantities in the oceans as sodium chloride. It is also a component of many minerals and is an essential element for human life.

Commercially, sodium metal is produced by electrolysis of sodium chloride. Sodium metal is widely used in the manufacture of chemicals and pharmaceuticals. Extremely dry solvents are important in chemistry for analytical as well as synthetic purposes. The drying process for hydrocarbons, ethers, and tertiary amines requires small pieces or wire of sodium. Large amounts of metallic sodium remain after the drying process, because the sodium reacts preferentially at its surface with the water of the solvent. These residues of sodium are normally destroyed by alcoholysis, which requires the addition of large amounts of alcohol. This operation is not only dangerous because of the formation of hydrogen gas: it is also considerably expensive.

Apparatus	Oil bath, a thick flattened glass rod, heating plate, glass vessel 10 cm in diameter and 20 cm in height with a wall thickness of 2 mm, stainless steel cylinder 9 cm in diameter and 15 cm in length equipped with a hole 2 mm in diameter in the bottom and a removable handle, sturdy wire 1.5 mm in diameter, stand, clamps, bosses, laboratory jack, safety glasses, laboratory coat, protective gloves.
Chemicals	Mineral oil, *n*-propanol (100 mL), sodium.
Attention!	Safety glasses and protective gloves must be worn at all times.

Caution!

Handling of sodium requires special care because sodium reacts explosively with water. The reaction with water is exothermic and produces hydrogen, which forms an explosive mixture on contact with air.

Never use sodium for drying halide-containing solvents! Large explosions can occur. In case of fire always use sand for extinguishing the fire. Always handle sodium in a well-ventilated hood having previously removed all chemicals from the hood. The sodium waste must always be covered with mineral oil.

Procedure

After the drying process, the residual sodium is recovered from the flask with a pair of tweezers. The sodium is placed in the glass vessel, and covered with mineral oil (Figure 1). The upper part of the vessel is attached to the stand with a clamp.

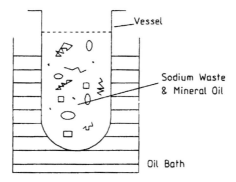

Figure 1   Melting process.

For the recovery of sodium, about 500 g of used sodium (sodium waste) are added to mineral oil (100–150 mL). The vessel is heated in an oil bath at about 100 °C. For this operation the oil bath is placed on a heating plate supported by a laboratory jack.

During the melting procedure, the sodium waste is compressed several times using the thick, flattened glass rod. After complete melting, the sodium forms a single layer below the mineral oil. The unrecyclable residues of higher density settle down at the bottom of the vessel. Now the steel cylinder with the removable handle (Figure 2) is lowered slowly into the melt. The mineral oil and the sodium enter through the 2 mm hole in the cylinder. This has to be done carefully so that the sodium does not flow over the edge into the cylinder (Figure 3).

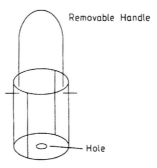

Removable Handle

Hole

Figure 2 Steel cylinder.

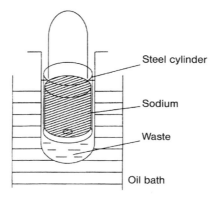

Steel cylinder

Sodium

Waste

Oil bath

Figure 3 Resolidification of sodium.

Finally, the oil bath is removed, and the mineral oil and the sodium in the steel cylinder are cooled to ambient temperature. Then the mineral oil is decanted and collected to use it for another batch. The steel cylinder with the solidified sodium is removed from the reaction vessel. After removing the handle, the sturdy wire is inserted through the hole in the bottom. The sodium is finally removed from the cylinder (Figure 4).

The recycled sodium is stored in a flask that contains high boiling petroleum. This sodium can be used again for drying solvents. The amount of recovered sodium is about 75-80%.

Waste Disposal The unrecovered sodium and the hydroxide waste are treated with *n*-propanol and then with water. The resulting sodium hydroxide solutions are collected and stored in a deposit for hazardous waste. The alcohol consumption is drastically decreased using this procedure.

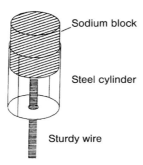

Figure 4   Removal of the sodium block

Normally, 10 L of alcohol are needed to treat 500 g of sodium waste. With this procedure only about 100 mL of alcohol are needed for decomposing the waste that contains the unrecovered sodium. In addition, the expenses for the deposit of waste are reduced considerably by minimizing the amount of waste.

Reference

1 B. Hübler-Blank, M. Witt, H.W. Roesky,
  J. Chem. Educ. **1993**, *70*, 408.

**Part V:
Special Topics**

34

Air and Nitrogen – A Light Bulb Turns On and Off

Hans-Dieter Barke

Teachers like to demonstrate the phenomenon of combustion by burning iron wool in air. Students are shown that there is an increase in mass due to the formation of iron oxide. One needs to consider that students never see burning metals in their everyday life – so they will argue that metals burn in chemistry laboratories but not in their environment.

Taking a light bulb, showing the glowing of tungsten, and demonstrating the destruction of the wire after producing a little hole in the glass of the bulb seems much more interesting for students. After they have discussed the glowing and burning of the metal wire in a light bulb, one can then burn other metals like iron or magnesium. One can compare the masses before and after burning, and then introduce the law of conservation of mass for chemical reactions.

After opening the bulb and destroying the metal wire, students may develop a hypothesis according to the role of the air – the role of oxygen and nitrogen. One possible proof of the hypothesis is taking liquid nitrogen, dipping the tungsten wire of the opened bulb into the liquid nitrogen, and turning the light on: the tungsten wire glows very brightly in the evaporating nitrogen. The students will never forget this interesting phenomenon – they now know much more about the combustion of metals and the construction of a light bulb than before.

It is also possible to connect these experiences to forensic chemistry. The position of the small hole is very interesting for forensic scientists, because they can determine the direction from which the fire originated. The presence of tungsten oxide on the wire or inside the bulb is used to determine whether the bulb was on or off during an accident [1].

Apparatus	Two light bulbs, beaker (500 mL), electric cable and plug, burner, crucible tongs, safety glasses, protective gloves.
Chemicals	Liquid nitrogen.

Attention!   To reduce risk of electric shock, unplug the power supply before directly handling the light bulb. Safety glasses and protective gloves must be worn at all times.

Caution!   You are working with high voltage. Liquid nitrogen can cause severe burns on the skin.

Procedure   (a) Plug one bulb in to the power supply and switch on to light the bulb. Heat the surface of the bulb and melt the glass with the hot flame of the burner until you hear a short sound: the bulb has a little hole. Wait.

   (b) Remove the glass from the second bulb by heating it, dipping it into cold water, and removing the glass using crucible tongs. Half-fill the beaker with liquid nitrogen, immerse the metal wire of the bulb in the liquid (using a stand) and wait one minute. Connect the bulb by cable and plug. Then switch on the electric current. Let the nitrogen evaporate and observe.

Observation   (a) White clouds appear and precipitate onto the inside of the glass (Figure 1), the light of the bulb goes out, and the metal wire is destroyed.

Figure 1   White film on inside of light bulb.

(b) The metal wire glows very brightly (Figure 2), the liquid nitrogen evaporates, and as soon as the air reaches the metal wire the light goes out and the metal wire is destroyed (Figure 3).

Figure 2   Open light bulb glowing brightly in beaker of liquid nitrogen.

Figure 3   Light bulb after exposure to air.

Explanation

The light bulb is filled with the noble gas argon so that the glowing wire cannot oxidize. As soon as air is present (instead of the argon), the reaction of tungsten and oxygen starts to form white tungsten trioxide:

$$W + 3/2 O_2 \rightarrow WO_3$$

Immersion in pure liquid nitrogen also protects the glowing wire from the air. The wire is glowing but not burning – just like in the argon atmosphere. If the nitrogen evaporates, air will cause the metal-oxygen reaction.

Reference

1 B. Rohrig, *Light bulb demos with a forensic twist*. In: Purdue University, 19th Biennial Conference on Chemical Education **2006**, Program Book, p. 266.

35

Limewater and CO₂ – A Light Bulb Turns On and Off

Hans-Dieter Barke

The reaction of carbon dioxide with limewater shows the well-known »milky« precipitation of white calcium carbonate. A lesser-known fact is that on further reaction with carbon dioxide, the white precipitate redissolves. Through this reaction one learns that in tap water or mineral water »soluble calcium« exists as a solution of calcium bicarbonate, i.e. $Ca^{2+}_{(aq)}$ ions and $HCO_3^-{}_{(aq)}$ ions. Insoluble calcium carbonate can be dissolved not only by strong acids, but also in weak carbonic acid solution. The natural cycle of CO_2 also works in this way. That is, the oceans of the world dissolve CO_2 from the air and lower the concentration of that »greenhouse gas«. Limestone can also be found in containers like water boilers or tea-kettles where tap water is repeatedly boiled. Even big stalagmites and stalactites are formed by calcium bicarbonate solution from which calcium carbonate precipitates out in dripstone caverns around the world.

Apparatus   Beaker (600 mL), test tubes, magnetic stirrer, conductivity tester (220 V glowing lamp) and pH meter (see Figure 1).

Figure 1 Experimental set-up.

Chemicals	Limewater and carbon dioxide gas.

Procedure   The beaker is one-third filled with limewater and placed on the magnetic stirrer, which is already turned on. The glass electrode of the pH meter is immersed and the pH value is measured; the conductivity meter is also immersed and the electrical conductivity (el.con.) is measured. Carbon dioxide gas is introduced until the initially formed white precipitate fully disappears [1].

Observation   (a) The pH meter shows that the pH of the limewater is 12.4, el.con. = 210 mA, and the bulb is glowing (Figure 1).

(b) On bubbling carbon dioxide through the limewater, a milky suspension appears, the pH value decreases and reaches pH 7, el.con. = 60 mA, the filament of the light bulb goes out.

(c) On adding more and more gas, the white precipitate completely disappears; the pH value decreases to pH 6, el.con. = 170 mA, and the light is on again.

Explanation

(a) The ions of lime water are $Ca^{2+}_{(aq)} + 2\ OH^-_{(aq)}$ pH 12.4; light turns on.

(b) $CO_{2\,(g)} + H_2O_{(l)} \rightleftarrows CO_{2\,(aq)}$

$OH^-_{(aq)} + CO_{2\,(aq)} \rightleftarrows HCO_3^-{}_{(aq)}$

$OH^-_{(aq)} + HCO_3^-{}_{(aq)} \rightleftarrows H_2O_{(l)} + CO_3^{2-}{}_{(aq)}$

$Ca^{2+}_{(aq)} + CO_3^{2-}{}_{(aq)} \rightleftarrows CaCO_{3\,(s,\ white)}$ pH 7; light turns off.

c) $CaCO_{3\,(s)} + H_2O_{(l)} + CO_{2\,(aq)} \rightleftarrows Ca^{2+}_{(aq)} + 2HCO_3^-{}_{(aq)}$

$H_2O_{(l)} + CO_{2\,(aq)} \rightleftarrows H^+_{(aq)} + HCO_3^-{}_{(aq)}$ pH 6; light turns on

$CO_{2\,(aq)} \rightleftarrows CO_{2\,(g)} + H_2O_{(l)}$

Reference

1 H.-D. Barke, A. Hazari, S. Yitbarek, *Misconceptions in Chemistry – Adressing Perceptions in Chemical Education*. Springer, Heidelberg, **2009**.

36

Template Synthesis of the Macrobicycle Dinitrosarcophagine

Jack M. Harrowfield and George A. Koutsantonis

The focus on sustainable chemistry in recent years reflects chemists' long-standing desire to provide efficient and cheap processes, with minimal environmental impact [1], for the synthesis of materials. The proving ground for these fundamental challenges is often the research laboratory, where the emphasis has been on benign solvents [2], efficient catalytic processes [3, 4] and more efficient energy sources [5]. A less well-trodden path is the one where the focus is on atom efficiency, because it is fundamentally more difficult to control both the kinetics and the energetics of a particular reaction, although some have tried [6].

Template processes have long been recognized as important in chemical syntheses, although they are commonly associated with reactions in which a metal ion acts as the »template« about which are assembled, appropriately juxtaposed, the constituents of a desired product. Much recent work has demonstrated the generality and wide applicability of the »template effect« [7]. The formation of macro(poly)cyclic molecules from acyclic, polyfunctional precursors is a well-known case where an initial intermolecular reaction is (desirably) followed exclusively by intramolecular events and where template control can be very effective, as well as being considerably more convenient to apply to large scale syntheses than alternative processes such as the use of »high-dilution« methods [8]. An example of a metal ion template reaction which can be readily conducted in the laboratory on any scale between ~0.5 g and 300 g of reactant (with larger amounts, temperature control for the exothermic reaction can become a problem in simple apparatus) is the synthesis of the Co(III) complex of the macrobicycle »dinitrosarcophagine«, (»sarcophagine« = »sar« = 3,6,10,13,16,19-hexa-azabicyclo[6.6.6]icosane) [9, 10]. Here, three 1,2-ethanediamine molecules bound to Co(III) are caused to react with six molecules of formaldehyde and two of nitromethane to give a single species in 70–80% yield, while the reaction of the same constituents but with Co(III) absent provides no detectable amount of dinitrosarcophagine. A multi-step mechanism can be postulated for the reac-

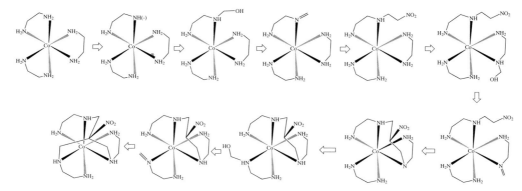

Scheme 1 A plausible mechanism for the formation of the Co(III) complex of 1,8-dinitrosarcophagine [3]. The steps shown are repeated for the capping of the second trigonal face of the reactant.

tion (Scheme 1), and it is clearly important to use an inert metal ion to preserve the octahedral geometry throughout the reaction, as labile metal ions do not function as effective templates. Dinitrosarcophagine and its complexes can be used as precursors to a wide variety of cage amine complexes [11], including some of biological and medical importance [12, 13].

Preparation of dinitrosarcophagine

Apparatus 4 L three-necked round-bottomed flask, magnetic stirrer, ice bath, retort stand and clamp, magnetic stirring bar, thermometer, safety glasses, laboratory coat, and protective gloves.

Chemicals $[Co(en)_3]Cl_3 \cdot 3H_2O$ (where en = 1,2-ethanediamine), $MeNO_2$, HCHO (37% aqueous solution), NaOH, HCl.

Attention! Care must be taken with solutions of caustic soda. Gloves and safety glasses must be worn.

Procedure A solution of $[Co(en)_3]Cl_3 \cdot 3H_2O$ (300 g) and $MeNO_2$ (180 mL) in a mixture of water (500 mL) and aqueous HCHO (37% solution; 700 mL) is cooled with an ice bath to 5 °C. (A period of stirring at

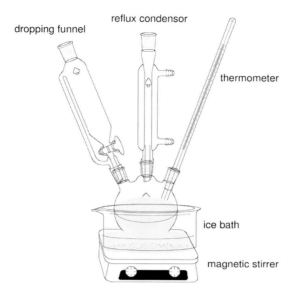

dropping funnel

reflux condensor

thermometer

ice bath

magnetic stirrer

Figure 1 Apparatus for the reaction. The size selected for the flask and dropping funnel is dependent on the scale of the reaction. The reaction mixture must be stirred vigorously using an appropriately sized magnetic stirring bar.

room temperature is necessary to ensure complete dissolution of the MeNO$_2$.) With vigorous stirring and continued cooling, a 5 °C solution of NaOH (120 g) in water (750 mL) is added all at once. The initially orange solution turns dark violet-brown within 60 s and the temperature rises rapidly to ~35 °C. If the reaction is not efficiently cooled in an ice bath it becomes violent, vigorous effervescence occurs, and an intractable brown tar is produced. After 15 min, the temperature drops to ~20 °C and the reaction is then quenched by the addition of concentrated HCl (1 L). With further cooling, a crystalline orange precipitate begins to form. After 2 h, this is collected and washed, first with cold HCl (1 mol L^{-1}) and then with ethanol, and allowed to dry under suction on the filter. Yield: 296 g (70%). The complex can be recrystallized from hot, dilute HCl by cooling with ice to give long, orange needles.

Characterization data: ^{1}H NMR (100 MHz, 36% DCl): δ 3.0–4.2 (complex asymmetric pattern, 24H, at least 9 C–H resonances), 8.10 (br s, 6H, N–H) ppm. ^{13}C NMR (15.04 MHz, D$_2$O): δ 20.0, –10.8, –14.4 ppm. IR (KBr, cm^{-1}): $\bar{v}$ 1555 (s), 1353 (s). UV (0.1 mol L^{-1} HCl):

λ max (lg ε) = 473.5 (146), 343.5 (124) nm. Additional spectroscopic characteristics of the complex are given in Ref. [9]. Calculated (found) for $[Co(C_{14}H_{30}N_8O_4)]Cl_3 \cdot H_2O$: C, 30.15 (30.3); H, 5.79 (5.6); Cl, 19.07 (19.1); Co, 10.57 (10.6); N, 20.09 (20.0) %.

Waste Disposal The waste solution from this synthesis is an aqueous filtrate containing Co complexes (only some of which have been identified) [9, 10], NaCl, HCl and presumably some tris(hydroxymethyl)nitromethane (precursor to the well-known buffer component »tris« = tris(hydroxymethyl)aminomethane). To recover the relatively precious Co, the simplest procedure is to allow the solution to evaporate in air, and then to roast the residue to recover the metal oxide.

References

1 J.H. Clark in *Green Separation Processes*, C.A.M Afonso (Ed.), Wiley, **2005**, 3–18.

2 (a) J.B.F.N. Engberts in *Methods and Reagents for Green Chemistry: An Introduction*, P. Tundo, A. Perosa, F. Zecchini (Eds.), Wiley, **2007**, 159–170; (b) N.V. Plechkova, K.R. Seddon, in *Methods and Reagents for Green Chemistry: An Introduction*, P. Tundo, A. Perosa, F. Zecchini (Eds.), Wiley, **2007**, 105–130.

3 J.M. Thomas, R. Raja, *Catal. Today* **2006**, *117*, 22.

4 M. Alcalde, M. Ferrer, F.J. Plou, A. Ballesteros, *Trends in Biotech.* **2006**, *24*, 281.

5 J-M. Leveque, G. Cravotto, *Chimia* **2006**, *60*, 313.

6 J. Xiang, A. Orita, J. Otera, *Angew. Chem.* **2002**, *114*, 4291; *Angew. Chem. Int. Ed.* **2002**, *41*, 4117.

7 Comprehensive reviews of template reactions, including that by D.H. Busch showing an historical perspective, are given in the two-volume publication of C.A. Schalley, F. Vögtle, K.H. Dötz (Eds), *Templates in Chemistry, I and II; Topics in Current Chemistry*, Vols 248, 249, Springer Verlag, Berlin, **2004**, **2005**.

8 B. Dietrich, P. Viout, J-M. Lehn, *Macrocycle Chemistry*, VCH, Weinheim, **1993**.

9 R.J. Geue, T.W. Hambley, J.M. Harrowfield, A.M. Sargeson, M.R. Snow, *J. Am. Chem. Soc.* **1984**, *106*, 5478.

10 G.A. Bottomley, I.J. Clark, I.I. Creaser, L.M. Engelhardt, R.J. Geue, K.S. Hagen, J.M. Harrowfield, G.A. Lawrance, P.A. Lay, A.M. Sargeson, A.J. See, B.W. Skelton, A.H. White, F.R. Wilner, *Aust. J. Chem.* **1994**, *47*, 143.

11 See, for example, J.M. Harrowfield, G.A. Koutsantonis, H.-B. Kraatz, G.L. Nealon, G.A. Orlowski, B.W. Skelton, A.H. White, *Eur. J. Inorg. Chem.* **2007**, 263 and references therein.

12 A.M. Sargeson, *Coord. Chem. Rev.* **1996**, *151*, 89.

13 S.D. Voss, S.V. Smith, N. DiBartolo, L.J. McIntosh, E.M. Cyr, A.A. Bonab, J.L.J. Dearling, E.A. Carter, A.J. Fischman, S.T. Treves, S.D. Gillies, A.M. Sargeson, J.S. Huston, A.B. Packard, *Proc. Nat. Acad. Sci. US* **2007**, *104*, 17489.

37

Cyclic Molecular Aluminophosphate: [Me$_2$Al(μ-O)$_2$P(OSiMe$_3$)$_2$]$_2$

Jiri Pinkas

Aluminophosphate molecular sieves are related to well-known aluminosilicate zeolites through the isoelectronic relationship among the AlPO$_4$, [AlSiO$_4$]$^-$, and (SiO$_2$)$_2$ formula units. They possess three-dimensional network structures composed of regularly alternating AlO$_4$ and PO$_4$ tetrahedra connected through all vertices by sharing the bridging oxygen atoms. These frameworks contain precisely defined pores and channels of dimensions up to 12 Å. The traditional synthetic route relies on a hydrothermal treatment of a source of aluminum, such as pseudo-boehmite or Al(Oi-Pr)$_3$, with aqueous H$_3$PO$_4$ in the presence of template compounds such as alkylamine, quaternary ammonium salt or aminoalcohol that facilitate the formation of microporous structures. The most frequently used solvent is water, but nonaqueous media were also employed. Even when working under predominantly dry conditions, a minimum amount of water from aqueous H$_3$PO$_4$, solvents, and pseudo-boehmite is always present in these reaction mixtures and cannot easily be removed. Water acts as a catalyst and mineralizer and plays an important role in determining the nature of the resulting framework. The hydrothermal conditions limit the choice of the structure-directing guests to inert, polar, and water-soluble compounds. To study the formation of aluminophosphates under completely nonaqueous conditions, different aluminum and phosphate sources must be employed and milder and nonaqueous reactions must be devised as alternatives to the hydrothermal synthesis. One of the soft chemical routes for the preparation of microporous aluminophosphates is the building block strategy. Here, preformed organic-soluble clusters of aluminophosphate possessing reactive groups are expected to condense and form extended three-dimensional structures. Pore size, shape, and dimensionality would be governed by the building unit characteristics. One of the model compounds that can be thought to mimic the building units is [Me$_2$Al(μ-O)$_2$P(OSiMe$_3$)$_2$]$_2$, which contains a central Al$_2$P$_2$O$_4$ ring. This molecule possesses reactive groups at the aluminum and phosphorus centers and can

serve in developing rational nonaqueous methods leading to microporous aluminophosphate materials.

Preparation of [Me₂Al(μ-O)₂P(OSiMe₃)₂]₂

Apparatus   Dry box, vacuum–nitrogen manifold, 250 mL Schlenk flask, rubber
                  septa, magnetic stirrer, syringes (10 and 20 mL), cannula, cold trap,
                  sublimation apparatus, spatulas, safety glasses, laboratory coat, pro-
                  tective gloves.

Chemicals   OP(OSiMe₃)₃, AlMe₃, Me₂AlCl, dry deoxygenated toluene, THF, and
                  hexane, liquid nitrogen.

Attention!   Safety glasses and protective gloves must be worn at all times.

Caution!   Because of their flammability, care should be taken when transfer-
                  ring AlMe₃ or Me₂AlCl to avoid contact of their solutions with air. All
                  reactions should be carried out in a well-ventilated hood. Contact of
                  liquid nitrogen with the skin or eyes may cause severe frostbite in-
                  jury.

Procedure   All preparative procedures are performed under a dry nitrogen at-
                  mosphere using Schlenk and dry box techniques. Solvents are dried
                  over and distilled from Na/benzophenone under nitrogen and de-
                  gassed prior to use.
                      Method A. The synthesis is carried out in two steps. First, the
                  adduct OP(OSiMe₃)₃·AlMe₃ is prepared, and this is then thermally
                  decomposed and provides the final product in a dealkylsilylation re-
                  action. A 250 mL Schlenk flask is charged in a dry box with OP(OSi-
                  Me₃)₃ (12.6 mL, 40.0 mmol). Outside the dry box, toluene (150 mL) is
                  added by a cannula, and the resulting clear solution is cooled to 0 °C
                  by an ice bath. Trimethylalane (AlMe₃) (22 mL, 44 mmol, 2 mol L⁻¹
                  solution in toluene) is then added drop by drop by a syringe to the
                  stirred reaction mixture. After slow warming to room temperature,
                  all volatile components are removed under vacuum. OP(OSiMe₃)₃·
                  AlMe₃ (14.4 g, 40 mmol) is transferred in a dry box to a sublimation
                  apparatus and is then heated under dry nitrogen at atmospheric

pressure to 200 °C for 4 h. The adduct undergoes a dealkylsilylation and provides a white crystalline solid which condenses on the cold finger. This product is subsequently resublimed at 120 °C/0.007 mm Hg, affording 8.5 g (75%) of the cyclic product.

Method B. Me$_2$AlCl (0.46 g, 5.0 mmol) is added slowly to OP(OSiMe$_3$)$_3$ (1.57 g, 5.0 mmol) in THF (40 mL) in a 250 mL Schlenk flask. The reaction mixture is heated at reflux for 5 h, and all volatile components are then removed under vacuum. The solid residue is recrystallized from cold hexane yielding 1.20 g (81%) of cyclic aluminophosphate.

Characterization data: m.p. 95–98 °C. ^{1}H NMR (C$_6$D$_6$): δ –0.32 (s, 6H, AlCH$_3$), 0.18 (d, $^4J_{PH}$ = 0.3 Hz, $^1J_{CH}$ = 119.7 Hz, ^{13}C satellites, $^2J_{SiH}$ = 7.0 Hz, ^{29}Si satellites, 18H, SiCH$_3$) ppm. ^{13}C NMR (C$_6$D$_6$): δ –9.4 (br s, AlCH$_3$), 0.43 (X of AA′X, $^3J_{PC}$ = 1.9 Hz, $^1J_{SiC}$ = 60.4 Hz, ^{29}Si satellites) ppm. ^{29}Si NMR (C$_6$D$_6$): δ 23.4 (X of AA′X, $^2J_{POSi}$ = 4.7 Hz, $^1J_{SiC}$ = 60.6 Hz, ^{13}C satellites) ppm. ^{31}P NMR (C$_6$D$_6$): δ –29.3 (s) ppm. IR (KBr pellet, cm^{-1}): $\bar{\nu}$ 2967 m, 2929 m, 2892 w, 1422 w, 1259 s, 1219 s (P=O), 1189 w, 1130 s, 1065 vs, 852 vs, 763 m, 684 vs, 613 m, 555 w, 464 w. MS (EI): m/z 581 (M–CH$_3$$^+$, 100%).

Waste Disposal All solvents must be collected in a properly labeled waste bottle and submitted for certified disposal. Metallic Na left after solvent drying must be carefully reacted in small portions with excess isopropanol. Traces of AlMe$_3$ and Me$_2$AlCl left in syringes and cannulas are neutralized with excess isopropanol. Used cannulas must be deposited in a needle disposal container. For an environmentally friendly disposal of sodium see the the contribution of H.W. Roesky in this book (Chapter 29).

Explanation

Trimethylalane (AlMe$_3$) is a strong Lewis acid and easily forms an adduct with OP(OSiMe$_3$)$_3$ by accepting a lone electron pair from phosphoryl oxygen. Upon heating, the solid adduct undergoes a dealkylsilylation by expelling tetramethylsilane molecules and provides a sublimable white crystalline solid. The driving force in this reaction is the formation of strong Al–O bonds. In the second method, the starting compounds react with elimination of trimethylchlorosilane. The molecular aluminophosphate is composed of a centrosymmetric $Al_2P_2O_4$ core in a chair conformation and peripheral reactive substituents.

References

1 J. Pinkas, D. Chakraborty, Y. Yang, R. Murugavel, M. Noltemeyer, H.W. Roesky, *Organometallics* **1999**, *18*, 523.

2 J. Pinkas, H. Wessel, Y. Yang, M.L. Montero, M. Noltemeyer, M. Fröba, H.W. Roesky, *Inorg. Chem.* **1998**, *37*, 2450.

3 D. Chakraborty, S. Horchler, R. Kratzner, S.P. Varkey, J. Pinkas, H.W. Roesky, I. Uson, M. Noltemeyer, H.-G. Schmidt, *Inorg. Chem.* **2001**, *40*, 2620.

4 Y. Yang, J. Pinkas, M. Noltemeyer, H.W. Roesky, *Inorg. Chem.* **1998**, *37*, 6404.

5 Y. Yang, J. Pinkas, M. Schäfer, H.W. Roesky, *Angew. Chem.* **1998**, *110*, 2795; *Angew. Chem. Int. Ed.* **1998**, *37*, 2650.

6 S.T. Wilson, B.M. Lok, C.A. Messina, T.R. Cannan, E.M. Flanigen, *J. Am. Chem. Soc.* **1982**, *104*, 1146.

7 S.T. Wilson, B.M. Lok, C.A. Messina, T.R. Cannan, E.M. Flanigen, *Intrazeolite Chemistry*; ACS Symp. Ser.: Washington, D. C., **1983**, *218*, 79.

8 R.E. Morris, S.J. Weigel, *Chem. Soc. Rev.* **1997**, 309.

9 M.E. Davis, R.F. Lobo, *Chem. Mater.* **1992**, *4*, 756.

38
Biogas

Stefan Zimmermann and Peter W. Roesky

The problem of global warming has added urgency to the search for renewable energy sources. As a consequence, biogas plants were established to produce methane from bio waste. Our aim was to develop a »biogas plant« which can be constructed from very rudimentary equipment to demonstrate the basic function of such a plant for a broader audience.

The anaerobic digestion of biodegradable waste can be divided into three steps. First, a hydrolysis of macromolecular compounds such as proteins, carbohydrates and fat leading to amino acids, disaccharides and fatty acids takes place. In the second step, these compounds are further converted by enzymes not only to acids and alcohols, but also to hydrogen and carbon dioxide. At the same stage, nitrogen and sulfur compounds are converted into ammonia and hydrogen sulfide. Finally, in the third step, under strictly anaerobic conditions, methane, carbon dioxide, and traces of ammonia, hydrogen and hydrogen sulfide are generated.

Biogas experiments [1, 2]

Biogas plant

Apparatus 2×1.0 L Erlenmeyer flasks, tapered cork stopper with hole, normal tapered cork stopper, small glass tube, 6-mm PVC hose, Pasteur pipette, thermometer, 3.0 L beaker, plastic tap, aquarium heater, safety glasses, laboratory coat, protective gloves.

Chemicals   Pig slurry, pond sediment, corn silage, sugar.

Attention!   Safety glasses and protective gloves must be worn at all times.

Caution!

Pig slurry and pond sediment may contain pathogens.

Methane is extremely flammable. Keep the container in a well-ventilated place away from sources of ignition. Take precautionary measures against static discharges. No smoking is allowed.

Procedure

The apparatus is set up according to Figure 1. Hands should be protected before the length of the pipette is adjusted to fit the equipment by breaking off the end. The pipette is more easily connected to the hose by warming the latter with a blow dryer beforehand. The aquarium heater must be used according to the instruction manual. The Erlenmeyer flask used for gas collection is best filled beforehand with water, closed with a rubber bung, and moved upside down to the beaker, which is also filled with water. Then, the bung is removed.

Place silage (250 g), which may be chopped up, into the other Erlenmeyer flask. To speed up the reaction, sugar (50 g) may be added. Then, water (250 mL) and pig slurry or pond sediment (250 mg) are added. The apparatus is closed with a bored rubber bung with a fitted glass tube which is attached to the hose, and then the aquarium heater is turned on. In a properly running apparatus, 1000 mL of gas should be obtained after 2 days using pig slurry and after 4 days using pond sediment.

To demonstrate the formation of methane, the Erlenmeyer flask used for gas collection is closed with a rubber bung, removed from the water bath, and turned the right way up. After 5–10 min, the gas can be lit with a candle at the opening of the Erlenmeyer flask. A steady flame is required to ignite the methane.

$$CH_4 + 2O_2 \rightarrow CO_2 + 2H_2O$$

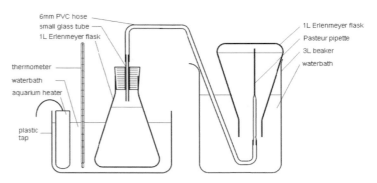

6mm PVC hose
small glass tube
1L Erlenmeyer flask
thermometer
waterbath
aquarium heater
plastic tap
1L Erlenmeyer flask
Pasteur pipette
3L beaker
waterbath

Figure 1 Experimental set-up.

Biogas plant for gas analysis

Apparatus 2 × 1.0 L Erlenmeyer flasks, tapered rubber bung with hole, normal tapered rubber bung, small glass tube, 6 mm PVC hose, Pasteur pipette, thermometer, 3.0 L beaker, plastic tap, aquarium heater, 2 × 250 mL gas washing bottles, 2 laboratory jacks, safety glasses, laboratory coat, protective gloves.

Chemicals Pig slurry, pond sediment, corn silage, sugar, copper(II) sulfate, barium hydroxide.

Attention! Safety glasses and protective gloves must be worn at all times.

Caution! Pig slurry and pond sediment may contain pathogens.

Methane is extremely flammable. Keep container in a well-ventilated place away from sources of ignition. No smoking is allowed. Take precautionary measures against static discharges.

Copper(II) sulfate pentahydrate is harmful if swallowed, as well as irritating to eyes and skin. It is very toxic to aquatic organisms and may cause long-term adverse effects in the aquatic environment. Do not breathe dust. This material and/or its container must be disposed of as hazardous waste. Avoid release to the environment. (Refer to special instructions safety data sheet.)

Barium hydroxide octahydrate is harmful by inhalation and if swallowed. It is corrosive and causes burns. In case of contact with eyes, rinse immediately with plenty of water and seek medical advice. Wear suitable protective clothing, gloves and eye/face protection. In case of accident or if you feel unwell, seek medical advice immediately (show label where possible).

Procedure The apparatus is set up according to Figure 2. Hands should be protected before the length of the pipette is adjusted to fit the equipment by breaking off the end. The pipette is more easily connected to the hose by warming the latter with a blow dryer beforehand. The aquarium heater must be used according to the instruction manual. The Erlenmeyer flask used for gas collection is best filled beforehand with water, closed with a rubber bung, and moved upside down to the beaker, which is also filled with water. Then, the bung is removed.

A saturated solution of $Ba(OH)_2$ is placed in one of the gas washing bottles. A solution of $CuSO_4$ (0.001 mol L^{-1}) is placed in the other gas washing bottle.

Place silage (250 g), which may be chopped up, into the other Erlenmeyer flask. To speed up the reaction, sugar (50 g) may be added. Then, water (250 mL) and pig slurry or pond sediment (250 mg) are added. The apparatus is closed with a bored rubber bung with a fitted glass tube which is attached to the hose, and then the aquarium heater is turned on. In a properly running apparatus, 1000 mL of gas should be obtained after 2 days using pig slurry and after 4 days using pond sediment.

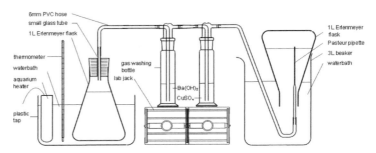

Figure 2 Experimental set-up for sample collection.

The formation of carbon dioxide can be shown by the reaction with barium hydroxide, which leads to barium carbonate in one of the washing bottles. The latter compound precipitates out of solution as a white solid.

$$Ba(OH)_2 + CO_2 \rightarrow BaCO_3\downarrow + H_2O$$

The formation of hydrogen sulfide is shown by the reaction with copper(II) sulfate. The reaction product, copper sulfide, is a black solid, which forms in the other washing bottle.

$$CuSO_4 + H_2S \rightarrow CuS\downarrow + H_2SO_4$$

To demonstrate the formation of methane, the Erlenmeyer flask used for gas collection is closed with a rubber bung, removed from the water bath, and turned the right way up. After 5–10 min, the gas can be lit with a candle at the opening of the Erlenmeyer flask. A steady flame is required to ignite the methane.

$$CH_4 + 2O_2 \rightarrow CO_2 + 2H_2O$$

Waste Disposal   Dispose of $BaCO_3$ in a manner consistent with federal, state, and local regulations.

References

1 H.G. Gassen, *Biol. Unserer Zeit* **2005**, *35*, 384.
2 D. Deublein, A. Steinhauser, *Biogas from waste and Renewable Resources*, Wiley-VCH, Weinheim, **2008**.

39

Preparation of Colloidal Cadmium Sulfide (CdS) Quantum Dot Nanoparticles

Dong-Kyun Seo

The size range of nanomaterials is comparable with some fundamental length scales in physics. The mean free path of an electron in a metal at room temperature, for example, is ~10–100 nm. The Bohr radius of photoexcited electron–hole pairs in semiconductors is ~1–10 nm. A quantum dot (QD) is a small particle of more or less isotropic shape with a diameter in the range from 1 to ~10 nm, matching the electron–hole pair radius of semiconductors. Some of the well-studied QDs include Au metal nanoparticles and ZnS, ZnSe, CdS, CdSe, PbSe, and InP semiconducting nanoparticles, among which CdSe QDs have been mostly studied. Size-controlled QDs can be prepared in several different ways, and in this experiment we prepare colloidal CdS QDs in oleylamine solution and characterize their optical properties.

In the zinc-blende–type structure of CdS, Cd and S atoms are bonded by four nearest neighbors and are periodically arranged with a cubic unit cell (Figure 1, left). The CdS unit cell, with a volume of 195 Å^3, contains 8 atoms, and thus the QDs with a diameter in range of 1–10 nm have approximately 20–20 000 atoms

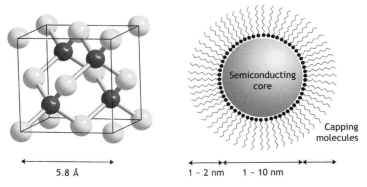

Figure 1 Left: Cubic unit cell of zinc-blende–type CdS (small spheres: Cd; large spheres: S). Right: Ideal structure of a semiconducting QD.

in their structure (Figure 1, right). The physical sizes and the number of atoms roughly match those of proteins. Understanding the electronic structure and optical properties of semiconducting QDs starts with considering the electronic nature of bulk semiconductors. First, we imagine a piece of a semiconductor as an extremely large molecule that contains almost Avogadro's number of atoms. Therefore, the electronic structure of the semiconductor exhibits an almost infinite number of molecular orbital (or more precisely crystal orbital) energy levels that are very close together and form energy bands. The bonding and antibonding interactions between atomic orbitals give rise to electron-filled valence bands and empty conduction bands (Figure 2, left). In CdS, the valence and conduction bands are formed mainly from S and Cd, respectively, and thus the compound has ionic configuration of $Cd^{2+}S^{2-}$. The energy gap between the top of the valence bands and the bottom of the conduction bands is called the »bandgap«. Light absorption by a semiconductor occurs when the energy of the incoming light is not less than the bandgap energy, inducing an electron excitation from one of the valence band energy levels to a conduction band energy level. Unlike molecules, however, absorption occurs continuously above the bandgap.

Light emission from a semiconductor can occur as a radiative relaxation process following the excitation. The nature of the excited state in a semiconductor is rather different from the molecular excited state because in a semiconductor there is strong delocalization of the excited electrons and the »holes« that are left behind by the excited electrons. The hole can be in fact thought of as

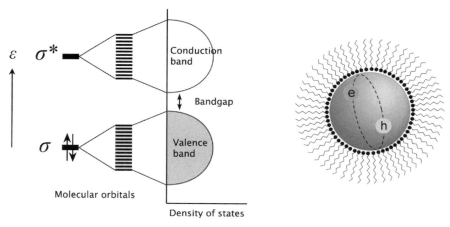

Figure 2 Left: Evolution of energy band structure of a semiconductor.
Right: Confinement of an electron-hole pair (exciton) in a semiconducting QD.

a particle with its own charge (+1) and effective mass. The electron and hole are often »bound« to each other for a short time via Coulombic attraction, and this pair is known as an »exciton«. The exciton can be considered a hydrogen-like system, and a Bohr approximation of the atom can be used in a classical picture to calculate the spatial separation of the electron-hole pair of the exciton by

$$r = \frac{\varepsilon h^2}{\pi m_r e^2} \tag{1}$$

where r is the radius of the sphere (defined by the three-dimensional separation of the electron-hole pair), ε is the dielectric constant of the semiconductor, m_r is the reduced mass, h is Planck's constant, and e is the electronic charge. For many semiconductors, the masses of the electron and hole are generally in the range 0.1–3 m_e (m_e is the mass of the electron). For typical dielectric constants, the calculation suggests that the electron-hole pair spatial separation can be as large as 10 nm for most semiconductors.

Now we consider what happens to the electronic structure of a semiconductor when the size of the semiconductor shrinks down to 1–10 nm diameter. Since the number of consisting atoms is much less in a quantum dot than in a bulk semiconductor ($\sim 10^{23}$), the broadening of the energy bands is much less, resulting in a larger energy separation between the valence and conduction bands. At the same time, because the physical dimensions of a QD can be smaller than the exciton diameter, the largest possible exciton diameter is no longer determined by Eq. (1) but by the size of the QD (»quantum confinement«; Figure 2 right). Here, the exciton in a QD is a good example of the »particle in a box«, and the lowest quantized energy of such confined excitons is generally taken as the absorption energy onset (bandgap) of the QD as in an early effective mass model (Brus model):

$$E_g(\text{QD}) = E_g(\text{bulk}) + \frac{h^2}{8R^2}\left(\frac{1}{m_e} + \frac{1}{m_h}\right) - \frac{1.8e^2}{4\pi\varepsilon_0\varepsilon R} \tag{2}$$

in which E_g is the optical energy gap of a QD or bulk solid, R is the QD radius, m_e and m_h are the effective masses of electron and hole in the solid, ε is the dielectric constant of the solid, and ε_0 is the vacuum permittivity. As the particle size decreases, the absorption onset shifts to higher energy (blue shifts), indicating an increase in optical energy gap. However, the experimental results have shown that this relationship does not hold very well, and the empirical E_g (QD)– R relationship can be used to estimate the size of the prepared QDs.

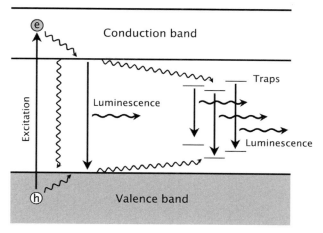

Figure 3 Various relaxation processes in a semiconductor. The small wavy arrows indicate nonradiative processes.

The most popular choices of QD materials for chemists are CdSe and CdS. Their bulk bandgaps are 1.7 eV for CdSe and 2.4 eV for CdS (corresponding to absorption onset at ~720 and ~520 nm, respectively), which means that their absorption and emission energies are tunable throughout the visible light region. Thus, CdS QDs have absorption spectra with absorption onsets at wavelengths theoretically from green through UV. QD size can be measured in several ways, including transmission electron microscopy (TEM), line-broadening of X-ray diffraction lines of QD powders, and electronic absorption spectroscopy, as given by Eq. (2) or related derivations.

The nature of the QD surface is critical for photoluminescence, since the probability of the radiative relation (i.e. quantum yield of the light emission) can be greatly reduced when the electrons and holes in the excited state are trapped by surface dangling bonds and undesired absorbent molecules. Electrons and holes in such trapped states can be recombined through nonradiative relaxation or luminescence that is substantially red-shifted from the exciton energy. »Surface passivation« is a well-known method for bulk semiconductors that can decrease the possibility of charge carriers residing in those traps. For QDs, surface passivation can be achieved by capping surface dangling bonds with suitable ligand molecules. In addition to the surface traps, there exist structural defects or impurities that act as a trap inside the QD particles. Because different traps give different light energies, luminescence from these random traps is characterized by a significantly broad peak in the photoluminescence spectrum of the

QDs. It appears in the longer wavelength region in comparison to the sharp peak from the exciton recombination. The internal traps can be eliminated only by optimization of reaction conditions.

In the synthesis of the CdS QDs, the capping molecules are mixed with the precursors of Cd and S in solution. During the formation of the QDs, the capping molecules restrict the growth of the particles by dynamically binding on the surface of the particles and remain bound to the QD surface at the end of the QD growth. Size tuning is possible by controlling relative concentrations of the reagents, reaction temperature and time. Another important aspect of surface capping is that the solubility of the QDs is controlled by their chemical characteristics. For example, a nonpolar end of a capping molecule allows dissolution of the QDs in nonpolar solvents, resulting in a colloidal QD solution, and provides a mean of purifying QDs through selective precipitation. The best-known synthetic route to synthesizing CdSe or CdS QDs is based on organometallic precursors, and careful control of reaction conditions allows us to produce QDs that are quite homogeneous in size (<5% standard deviation in diameter for a given batch) and tunable in size from 2 to 8 nm. Such a degree of size distribution provides a bell-shaped emission peak with its full-width-half-maximum (FWHM) of ~30 nm for CdSe QDs. The estimated FWHM for a single CdSe QD is ~16 nm. Post-preparative size fractionation is also often used to narrow the size distribution by exploiting the difference in solubility of smaller and larger particles. In this experiment, we employ a less hazardous synthetic method that operates at much lower temperatures. Oleylamine is used as a solvent for $CdCl_2$ and P_2S_5, and it also serves as the capping molecule of the CdS QD product. The amine group binds to the surface of the CdS QDs and the QDs are freely soluble in hexanes because of the nonpolar oleyl group.

Apparatus   Two laboratory ovens (or block heaters for vials), centrifuge, 4 crimp-cap thick-walled reaction vials (~15 mL), 4 aluminum crimp-caps with Teflon-lined rubber septum, 1 vial crimper and decapping plier, 2 centrifuge tubes, 3 graduated cylinders (20 mL), microbalance, 2 spatulas, UV lamp (λ_{ex} = ~310 nm), UV-Vis spectrometer, fluorimeter, silica cuvettes, safety glasses, laboratory coat, and protective gloves.

Chemicals   Oleylamine (*cis*-1-amino-9-octadecene); technical grade (70%), cadmium chloride ($CdCl_2$), diphosphorus pentasulfide (P_2S_5), ethanol, hexanes.

Attention!   The centrifuge should be well balanced to avoid a serious injury and students are required to ask help from the instructor in balancing the sample weights. Handling of the chemicals and products should be carried out in a well-ventilated fume hood. Wear safety glasses and latex gloves throughout the experiment.

Caution! P_2S_5 reacts with moisture relatively slowly to release H_2S, a toxic gas with a foul smell. The apparatus or material that has been in contact with P_2S_5 and its solution should be transferred immediately into a fume hood. Cadmium is a toxic heavy element and should be handled and collected in a container for heavy toxic metals.

Procedure   Place P_2S_5 (0.3 mmol) and oleylamine (10 mL) in a reaction vial. Seal the vial tightly with a septum aluminum cap by using a vial crimper. In another vial, place $CdCl_2$ (0.06 mmol) and oleylamine (10 mL), and seal the vial.

Transfer the vials to an oven and heat them at 110 °C for 30 min with occasional shaking in order to dissolve the precursors. After cooling down to room temperature, the solutions should remain colored but clear.

Open the vials with a decapping plier and mix the two precursor solutions well. Divide the mixture into two vials and seal the vials tightly. Heat one mixture solution at 110 °C and the other at 180 °C for 1 h.

After cooling down the vials to room temperature, transfer each solution into a centrifuge tube. Add ethanol (~20 mL) to each tube. Close the cap of the tube and shake the tube well. This will result in flocculation and precipitation of the QD nanoparticles. Separate the precipitate from the solution by centrifugation at ~2500 rpm. Decant the supernatant solution carefully. Take a small amount of the precipitate and disperse it in hexanes (~10 mL). The solutions obtained should be optically clear. The photoluminescence of the solutions should be apparent under a UV lamp ($\lambda_{ex} = \sim310$ nm).

Obtain UV-Vis absorption and photoluminescence spectra of the two QD products in hexanes. Your TA will help you dilute the solutions properly for your absorption and photoluminescence ($\lambda_{ex} = 310$ nm) measurements.

Waste Disposal   Because of the toxicity of the cadmium, all the apparatus and waste chemicals should be disposed of in the container for toxic heavy metals. The Cd precursor solution can be rinsed off with acetone and the product solutions with hexanes.

Explanation

1. Provide probable chemical equations for the steps 2 (dissolution of $CdCl_2$ and P_2S_5) and 3 (formation of CdS). Assume that P_2S_5 reacts quantitatively with a primary amine to provide $[HS]^-$ as the reactive S species in the precursor solution. It is also reasonable to assume that Cd^{2+} ions are coordinated by six amine molecules under the reaction conditions.

2. Locate the first exciton peak maximum in the absorption spectra. Discuss the differences between the two spectra based on the particle sizes. Higher reaction temperatures allow faster particle growth and thus larger QD sizes. Quantitative estimation of the QD sizes may be carried out by employing the empirical (QD)– relationship reported in Ref. [8].

3. Identify the emission peak from the exciton in the photoluminescence spectra. Examine for the possible presence of internal traps in relaxation of the excitons. Discuss the differences between the two spectra based on the particle sizes. It is known that bulk CdS materials do not exhibit recognizable photoemission. Based on the quantum confinement effect, explain why the CdS QDs have a superior photoluminescence property in comparison to bulk CdS.

References

Adopted synthetic method
1 N. Iancu, R. Sharma, D.-K. Seo, *Chem. Commun.* **2004**, 2298.
2 Q. Wang, D.-K. Seo, *Chem. Mater.* **2006**, *18*, 5764.

Other synthetic methods
3 C.B. Murray, D.J. Norris, M.G. Bawendi, *J. Am. Chem. Soc.* **1993**, *115*, 8706.
4 Z.A. Peng, X. Peng, *J. Am. Chem. Soc.* **2002**, *124*, 3343.
5 J.J. Li, Y.A. Wang, W. Guo, J.C. Keay, T.D. Mishima, M.B. Johnson, X. Peng, *J. Am. Chem. Soc.* **2003**, *125*, 12567.

Optical properties of QDs
6 S.V. Gaponenko, *Optical Properties of Semiconductor Nanocrystals*, Cambridge University Press, New York, **1998**.
7 L.E. Brus, *J. Chem. Phys.* **1984**, *80*, 4403.
8 W.W. Yu, L. Qu, W. Guo, X. Peng, *Chem. Mater.* **2003**, *15*, 2854.

Overview
9 A.P. Alivistos, *Science* **1996**, *271*, 933.
10 C.B. Murray, C.R. Kagan, M.G. Bawendi, *Annu. Rev. Mater. Sci.* **2000**, *30*, 545.
11 C.J. Murphy, *Anal. Chem.* **2002**, *74*, 520A.

40

Dendrimer Construction: [CpFe]⁺-Induced Formation of the Polyallyl Arene Core and Dendron, and Iteration using Hydrosilylation and Williamson Reactions

Jaime Ruiz, Catia Ornelas, Rodrigue Djeda and Didier Astruc

The divergent synthetic strategy enabling large dendrimers to be synthesized was disclosed by the groups of Tomalia [1], Newkome [2] and Denkewalter [3] in the early 1980s subsequent to a report of an iteration of polyamine synthesis by Vögtle [4] in 1978. (The alternative convergent synthesis, limited to smaller dendrimers but with fewer defects, was published in 1990 by the groups of Fréchet and Miller [5, 6].)

In January 1979, the iterative sequential deprotonation–alkylation of the sandwich complex $[FeCp(\eta^6\text{-}C_6Me_6)][PF_6]$ was reported to quantitatively yield $[FeCp(\eta^6\text{-}C_6Et_6)][PF_6]$ after a spontaneous reflux of about 1 min upon reaction with t-BuOK and MeI in THF [7]. The [CpFe]⁺-induced synthesis of star-shaped molecules with a variety of substituents is readily available in this way (Scheme 1).

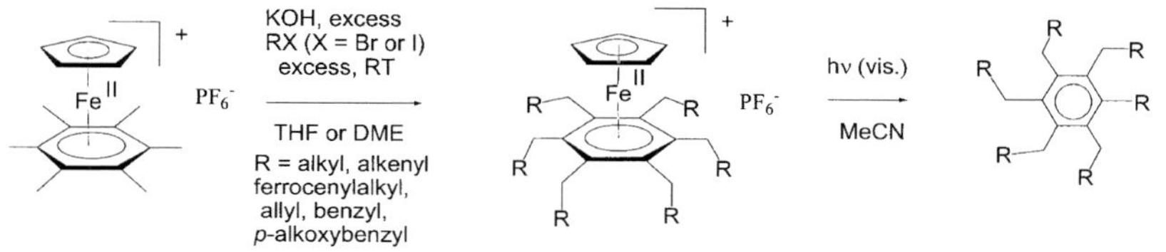

Scheme 1

The mono-deprotonation complex and mono-methylated intermediate complex were isolated, confirming the mechanism involving six deprotonation–alkylation sequences [8]. This was taken into account by the fact that the pK_a of C_6Me_6 in DMSO drops from 44 to 29 upon complexation in $[FeCp(\eta^6\text{-}C_6Me_6)][PF_6]$ [9]. These spectacular and potentially very useful organometallic reactions were the start of dendrimer chemistry in our group [10], especially since the nona-allylation of the mesitylene complex $[FeCp(\eta^6\text{-}C_6H_3Me_3)][PF_6]$ using t-BuOK (or KOH) and allyl bromide was shown to proceed analogously, although more smoothly, but also quantitatively at room temperature (Scheme 2) [11].

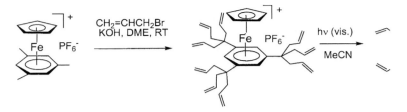

Scheme 2

Finally, the simple but extremely useful phenol dendron p-HO-C_6H_4-C(allyl)$_3$ could also be synthesized in 50% yield in a one-pot reaction from the p-ethoxytoluene complex [FeCp(η^6-p-EtO-C_6H_4Me)][PF$_6$], t-BuOK and allyl bromide under very mild conditions (Scheme 3) [12, 14].

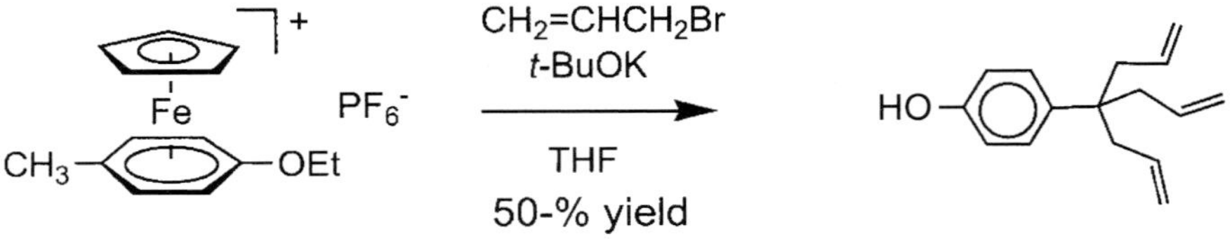

Scheme 3

These polyfunctionalizations represent excellent examples of the »Umpolung« of aromatic reactivity [15]. The [CpFe]$^+$-induced arene polyallylation reactions are detailed in Scheme 4 below together with the two reactions involved in the iteration from one dendrimer generation to the next (for instance, below

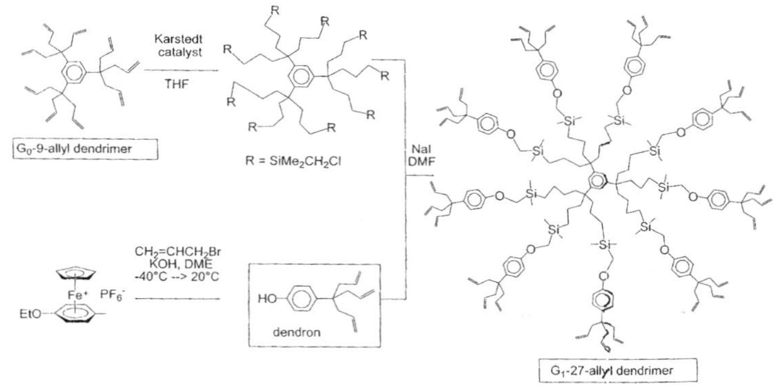

Scheme 4

from G_0-9-allyl to G_1-27-allyl): the hydrosilylation and the Williamson reaction. Experimental details are provided here for these four reactions.

The hydrosilylation reaction uses dimethylchloromethylsilane and is catalyzed by the Karstedt catalyst. Nucleophilic substitution of the terminal chloride group by the dendronic phenolate in DMF is facilitated and catalyzed by NaI, initial nucleophilic substitution of the chloride by iodide followed by nucleophilic substitution of iodide by the phenolate being faster than the direct substitution of chloride by the phenolate. The dendritic nona-allyl core, resulting from the $[CpFe]^+$-induced nona-allylation of mesitylene, and the phenol dendron have been shown to iteratively lead to high dendrimer generations including giant ones with 3^{n+2} terminal tethers (n = number of the dendritic generation G_n) by multiplication by three of the branch number at each generation [13].

Some experiments in the creation of dendrons

p-Triallylmethylphenol dendron

Apparatus	Stirring plate, thermometer, bench-top dewar (500 mL), three-necked flask, stirring bar, cannula, rubber septum, safety glasses, protective gloves and laboratory coat.
Chemicals	Tetrahydrofurane (THF) was purified before use by distillation from sodium/benzophenone ketyl under N_2. Pentane, diethyl ether, potassium *tert*-butoxide (*t*-BuOK), allyl bromide, aqueous solutions of hydrochloric acid (6 mol L^{-1}) and sodium sulfate were purchased from a laboratory chemical supplier and used as received.
Attention!	All reactions should be carried out in a well-ventilated hood; safety glasses, laboratory coat and protective gloves must be worn at all times.
Caution!	THF/sodium is highly flammable. Contact with air must be avoided. THF/sodium should be handled under N_2 or argon because of its toxicity and volatility. Care should be taken to avoid inhalation of or direct contact with dichloromethane and allyl bromide. *t*-BuOK and aqueous solutions of hydrochloric acid are irritant to eyes, respiratory system and skin.

Procedure   The compounds [FeCp(η^6-p-EtOC$_6$H$_4$Me)][PF$_6$] [12] (5 g, 12.4 mmo1) and t-BuOK (12.6 g, 112.5 mmol) are introduced into a flamed three-necked Schlenk flask under N$_2$, then 100 mL of freshly distilled THF are introduced by cannula into this flask at –50 °C. After stirring for 15 min, cold allyl bromide (20 mL, 0.232 mmol) is added at –50 °C. The reaction medium is allowed to warm up and further stirred at ambient temperature in the dark for one day. Then t-BuOK (12.6 g, 112.5 mmol) and allyl bromide (10 mL, 0.116 mmol) in THF (50 mL) are introduced at –50 °C, and the reaction mixture is allowed to warm up and further stirred for one more day at ambient temperature in the dark. This addition followed by one-day reaction is then repeated identically once more.

Finally, t-BuOK (9 g, 80 mmol) dissolved in 10 mL of THF is added at –50 °C, and the reaction mixture is allowed to warm to room temperature over 2 h and stirred for two more days under ambient conditions. The solvent is removed under vacuum, the solid residue is dissolved in 500 mL ether, and this solution is washed with 100 mL HCl (6 mol L^{-1}), dried over sodium sulfate and filtered. The solvent is removed under vacuum and the residue is chromatographed on a silica column using pentane/ether (95/5) as the eluent, which gives p-triallylmethylphenol as an off-white solid (1.71 g, 0.75 mmol, 50% yield). R_f 0.5 in CH$_2$Cl$_2$ (silica gel plate).

Characterization data: ^{1}H NMR (250 MHz, CDCl$_3$): δ 7.18 (d, 2H, CH_{arom}), 6.83 (d, 2H, CH_{arom}), 5.68.5.55 (m, 3H, CHCH_2), 5.07–5 (m, 6H, CHCH_2), 2.44 (d, 6H, CH_2) ppm. ^{13}C NMR (62.9 MHz, CDCl$_3$): δ 153.40 (C$_{quat.\ arom.}$COH), 137.60 (C$_{quat.\ arom.}$COH), 134.56 (CHCH$_2$), 127.73 (CH$_{arom.}$COH), 117.42 (CHC$\underline{H}_2$), 114.89 (C$\underline{H}_{arom.}$C), 42.59 (C$_{quat}$,–CH$_2$), 41.87 (CH$_2$) ppm. Calculated (found) for C$_{16}$H$_{20}$O: C, 84.16 (84.14), H, 8.83 (8.82)%.

Waste Disposal   Solvent-free aqueous layers are disposed of into the sewer system. Halogenated solvents and nonhalogenated solvents are collected separately in containers for disposal. Celite and silica are collected in powder containers for disposal.

[FeCp(η^6-1,3,5-tris(triallylmethyl)benzene] [PF$_6$]: the nona-allylation reaction

Apparatus Stirring plate, thermometer, bench-top dewar (500 mL), three-necked Schlenk flask, stirring bar, cannula, safety glasses, protective gloves and laboratory coat.

Chemicals 1,2-Dimethoxyethane (DME) is purified before use by distillation from sodium/benzophenone ketyl under argon. Pentane, allyl bromide, hexafluorophosphoric acid (HPF$_6$) 60% aqueous solution, sodium hydroxide and sodium sulfate are purchased from laboratory chemical suppliers and used as received.

Attention! All reactions should be carried out in a well-ventilated hood; safety glasses, laboratory coat and protective gloves must be worn at all times.

Caution! DME/sodium is highly flammable; avoid contact with air; DME/sodium should be handled under N$_2$ or argon. Care should be taken to avoid inhalation or direct contact with the solvents DME and dichloromethane because of their toxicity and volatility. Sodium hydroxide and aqueous HPF$_6$ solutions are highly irritant to the eyes, respiratory system and skin.

Procedure Using standard Schlenk techniques, [FeCp(η^6-mesitylene)][PF$_6$] [10–12] (5 g, 13.00 mmol) and finely crushed and dried KOH (33 g, 588 mmol) are introduced into a flamed three-necked flask under N$_2$. A mixture of DME (100 mL) and allyl bromide (51 mL, 180 mmol) in a Schlenk flask are flushed with N$_2$, then introduced by cannula into the three-necked flask at –50 °C. After warming up, the reaction medium is stirred at ambient temperature in the dark for two days, then brought to –50 °C, and a mixture of finely crushed and dried KOH (33 g) and cold allyl bromide (51 mL) is added. This reaction mixture is stirred for two days in the dark at ambient temperature, the solvent is removed under vacuum, CH$_2$Cl$_2$ (100 mL) is added, and the solid suspension is filtered over celite.

The soluble organic phase is washed with 50 mL of an aqueous solution of 5% HPF$_6$, dried over sodium sulfate and filtered. The solvent is removed under vacuum, and the brown waxy product is washed three times with 30 mL of pentane. The brown solid is dissolved in CH$_2$Cl$_2$, a flash chromatography is carried out on activated alumina, and the solvent is reduced to 5 mL under reduced pressure. Addition of 50 mL of diethyl ether to the solution precipitates the desired product [FeCp(η^6-1,3,5-tris(triallylmethyl)benzene][PF$_6$] as a yellow powder (6.31 g, 65%).

Characterization data: ^{1}H NMR (250 MHz, CDCl$_3$): δ 6.02 (s, C$_6$H$_3$), 5.89–5.65 (m, 9H, CHCH$_2$), 5.24–5.17 (m, 18H, CHCH_2), 4.94 (s, 5H, C$_5$H$_5$), 2.54 (d, 18H, CH_2) ppm. ^{13}C NMR (62.9 MHz, CDCl$_3$): δ 132.43 (CHCH$_2$), 120.36 (CHCH_2), 113.0 (C$_{quat.\ arom.}$), 76.69 (C$_5$H$_5$), 43.49 (C$_{quat.}$–CH$_2$), 42.62 (CH$_2$) ppm.

Waste Disposal Solvent-free aqueous layers are disposed of into the sewer system. Halogenated solvents and nonhalogenated solvents are collected separately in containers for disposal. Celite and silica are collected in powder containers for disposal.

Nonaallyl dendrimer core 1,3,5-tris(triallylmethyl)benzene: G$_0$-9-allyl

Apparatus Stirring plate, double-wall Schlenk flask, stirring bar, cannula, rubber septum, safety glasses, protective gloves and laboratory coat.

Chemicals MeCN, purified before use by distillation from P$_4$O$_{10}$ under N$_2$; PPh$_3$ and pentane, purchased from laboratory chemical supplier and used as received.

Attention All reactions should be carried out in a well-ventilated hood; laboratory coat, safety glasses and protective gloves must be worn at all times.

Caution! P$_4$O$_{10}$ is irritating to the eyes, respiratory system and skin.

Procedure In a double-wall Schlenk flask, the complex [FeCp{η^6-1,3,5-tris(triallylmethyl)benzene}][PF$_6$] (6.3 g, 8.43 mmol) and PPh$_3$ (2.13 g, 8.03 mmol) are dissolved in dry MeCN. The reaction mixture is degassed by flushing nitrogen for 5 min, then irradiated with visible light for 15 h under magnetic stirring and water cooling of the exterior wall of the Schlenk tube (20 °C). The solvent is removed under vacuum, and the dark purple product, [FeCp(MeCN)$_2$PPh$_3$][PF$_6$], is washed three times with pentane (30 mL). The pentane solution is evaporated under vacuum, and 1,3,5-tris(triallylmethyl)benzene is obtained as a white solid (3.06 g, 75%).

Characterization data: ^{1}H NMR (250 MHz, CDCl$_3$): δ 7.04 (s, 3H, C$_6$H$_{3arom}$), 5.55–5.45 (m, 9H, C*H*CH$_2$), 5.00–4.94 (m, 18H, CHC*H*$_2$), 2.42 (d, 18H, C*H*$_2$) ppm. ^{13}C NMR (62.9 MHz, CDCl$_3$): δ 144.38 (C$_{quat.\ arom}$), 134.71 (*C*HCH$_2$), 122.65 (CH$_{arom.}$), 117.47 (CH*C*H$_2$), 43.68 (C$_{quat}$,–CH), 41.98 (*C*H$_2$) ppm. Calculated (found) for C$_{36}$H$_{48}$: C, 89.93 (89.14), H, 10.06 (10.32)%.

Waste Disposal Solvent-free aqueous layers are disposed of into the sewer system. Halogenated solvents and nonhalogenated solvents are collected separately in containers for disposal. Celite and silica are collected in powder containers for disposal.

Hydrosilylation of the nona-allyl dendrimer core G$_0$-9-allyl

Apparatus Stirring plate, single-necked Schlenk flask, stirring bar, cannula, rubber septum, protective gloves, safety glasses, laboratory coat.

Chemicals Diethyl ether is purified before use by distillation from sodium/benzophenone ketyl under N$_2$; dichloromethane, dimethylchloromethylsilane and bis(divinyltetramethyldisiloxane)platinum are purchased from laboratory chemical suppliers and used as received.

Attention! All reactions should be carried out in a well-ventilated hood; laboratory coat, safety glasses and protective gloves must be worn at all times.

Caution! Sodium and diethyl ether are highly flammable, and contact with air must be avoided. Sodium and diethyl ether should be handled under N$_2$ or Ar. Care should be taken to avoid inhalation or direct contact with dichloromethane because of its toxicity and volatility.

Procedure *Synthesis of G$_0$-9-SiMe$_2$Cl.* Dry diethyl ether (20 mL), then, in a glove box, 10 drops of a commercial xylene solution (2%) of the Karstedt catalyst bis(divinyltetramethyldisiloxane)platinum and dimethylchloromethylsilane (1.520 g, 13.94 mmol) are successively added to the nona-allyl core (0.500 g, 1.04 mmol) G$_0$-9-allyl in a Schlenk flask. The reaction mixture is allowed to stir under argon for 12 h at room temperature, the solvent and excess of silane are removed under vacuum, and the yellow oily residue is dissolved in CH$_2$Cl$_2$ and filtered over a short column of silica in order to remove the catalyst. The solvent is then removed under vacuum, and a colorless oil of the nonachloromethylsilyl dendrimer G_0-9-$SiCH_2Cl$ is obtained (1.50 g; 98%).

Characterization data: ^{1}H NMR (CDCl$_3$, 300 MHz): δ 6.96 (Ar, s), 2.73 (CH_2Cl, s), 1.60 ($CH_2CH_2CH_2Si$, m), 1.09 ($CH_2CH_2CH_2Si$, m), 0.60 ($CH_2CH_2CH_2Si$, t), 0.06 ($SiMe_2$, s) ppm. ^{29}Si NMR (CDCl$_3$, 59.6 MHz): δ 3.85 ppm. ^{13}C NMR (100.6 MHz): δ 145.6 (substituted arene C), 121.4 (arene CH). 43.8 (quaternary benzylic C), 41.8 (ArCCH_2), 30.4 (SiCH_2Cl), 17.8 ($CH_2CH_2CH_2$), 15.8 (SiCH_2CH_2), –4.2 (SiCH_3) ppm.

Waste disposal Solvent-free aqueous layers are disposed of into the sewer system. Halogenated solvents and nonhalogenated solvents are collected separately in containers for disposal. Celite and silica are collected in powder containers for disposal.

Nucleophilic substitution of the chloride by the triallylmethyl phenolate dendron: synthesis of G_1-27- allyl from G_0-9-$SiCH_2Cl$

Apparatus Hot-plate stirrer, single-necked Schlenk flask, stirring bar, cannula, rubber septum, protective gloves, safety glasses, laboratory coat.

Chemicals Anhydrous *N,N*-dimethylformamide (DMF), methanol, dichloromethane, potassium carbonate, sodium iodide, celite and silica gel are purchased from laboratory chemical supplier and used as received.

Attention! All reactions should be carried out in a well-ventilated hood; safety glasses and protective gloves must be used at all times.

Caution! Care should be taken to avoid inhalation or direct contact with the solvents DMF, methanol and dichloromethane because of their toxicity and volatility. Potassium carbonate is irritating to the eyes, respiratory system and skin.

Procedure *Synthesis of G_1-27-allyl by reaction of G_0-9-$SiCH_2Cl$ with the triallylmethylphenolate dendron.* G_0-9-$SiCH_2Cl$ (0.500 g, 0.34 mmol), the triallylmethyl phenolate dendron (0.845 g, 3.70 mmol), K$_2$CO$_3$ (4.26 g, 30.9 mmol), and NaI (0.093 g, 0.62 mmoL) are introduced into a

Schlenk flask under N$_2$, then 50 mL of DMF are also transferred into the Schlenk flask via cannula, and the reaction mixture is stirred at 80 °C for 1 day. Then, the solvent is removed under vacuum, and the residue is dissolved in dichloromethane (20 mL) and filtered over celite. This solution is washed with an Na$_2$S$_2$O$_3$-saturated aqueous solution and dried over Na$_2$SO$_4$. After filtration and reduction of the solvent volume to 5 mL under reduced pressure, methanol (100 mL) is added in order to precipitate the product. The gummy residue obtained is dissolved in dichloromethane (5 mL) and submitted to chromatography on a column of silica gel (eluent: dichloromethane), which yields 0.801 g (73%) of the G$_1$-27-allyl dendrimer as a waxy solid.

Characterization data: ^{1}H NMR (CDCl$_3$, 400 MHz): δ 7.165 and 6.85 (*p*-C$_6$H$_4$, double d), 5.535 (–C*H*CH$_2$, m), 4.96 (–CHC*H$_2$*, m), 3.48 (SiC*H$_2$*, s), 2.40 (–C*H$_2$*–CHdCH$_2$, d), 1.63 (C*H$_2$*CH$_2$CH$_2$Si, m), 1.14 (CH$_2$C*H$_2$*–CH$_2$Si, m), 0.57 (CH$_2$CH$_2$C*H$_2$*Si, m), 0.03 (SiMe$_2$) ppm. ^{13}C NMR (100.6 MHz): δ 145.6 (substituted arene *C*), 137.20 (quaternary arene *C*), 127.10 (arene *C*H), 121.43 (arene *C*H), 117.40 (–*C*HCH$_2$,), 113.49 (arene *C*H), 60.11 (Si*C*H$_2$O), 42.58 (quaternary benzylic *C*), 41.96 (quaternary benzylic *C*), 17.86 (CH$_2$*C*H$_2$CH$_2$), 14.60 (Si*C*H$_2$CH$_2$), –4.2 (Si*C*H$_3$) ppm. Calculated (found) for C$_{207}$H$_{300}$O$_9$Si$_9$: C, 78.05 (78.25); H, 9.49 (9.50)%.

Waste Disposal Solvent-free aqueous layers are disposed of into the sewer system. Halogenated solvents and nonhalogenated solvents are collected separately in containers for disposal. Celite and silica are collected in powder containers for disposal.

Explanation

The basis of the »Umpolung« of arene reactivity in the 18-electron cationic sandwich complexes [FeCp(η^6-arene)]$^+$ is the strong electron-withdrawing effect of the 12-electron [CpFe]$^+$ group that is about equivalent to two nitro groups. This makes the arenes unreactive with electrophiles, acids and mild oxidants but reactive towards nucleophiles, bases and reductants, when they are engaged in such cationic iron complexes [7, 15]. Benzylic deprotonation becomes possible with bases such as *t*-BuOK or KOH. Even if these bases are not strong enough to completely deprotonate the iron complex in the benzylic position, the deprotonation can be rendered complete if an electrophile is added to the medium. The electrophile reacts with the deprotonated iron complex and therefore shifts the equilibrium towards the products. The deprotonated complex can in fact be represented as two mesomeric forms (the X-ray crystal structure shows a dihedral angle of 32° for the deprotonated hexamethylbenzene complex, justifying the structure of the cyclohexadienyl form shown below), but smooth reactivity with a very large range of electrophiles is explained by the zwitterionic form that has a minor weight in the structure) [8].

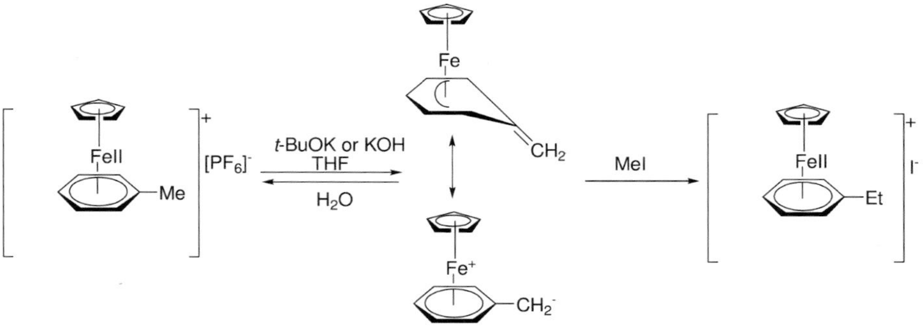

Scheme 5

After the deprotonation–alkylation sequence, the complex is again in its cationic form and contains other benzylic hydrogens that undergo other deprotonation–alkylation sequences. In the hexamethylbenzene complex, the hexabranch bulk considerably inhibits a second substitution on each benzylic carbon, but in the mesitylene and *para*-ethoxytoluene complexes, the exo-cyclic methyl group has no neighbor, and three deprotonation–alkylation (or allylation) sequences can easily substitute the three benzylic hydrogen atoms of the methyl group in one pot [11].

The four experimental procedures described here are the basis of the divergent dendrimer construction of polyallyl and other polyfunctional dendrimers derived from it. These polyallyl dendrimers have been synthesized up to G_9 and characterized by ^{1}H, ^{13}C and ^{29}Si NMR, transmission electron microscopy (after osmylation of the terminal double bonds), atomic force microscopy (which enables the thickness of the dendrimers to be measured on a mica plate in the condensed phase), elemental analysis and size exclusion chromatography (SEC) up to G_5. The NMR spectra show that each reaction is apparently quantitative within the accuracy of NMR (> 95%). MALTI TOF mass spectra recorded up to G_4 show the molecular peaks that are largely dominant for G_1 and G_2, but already minor for G_3, as defects intrinsic in the divergent synthesis appear. The SEC measurements show symmetrical peaks up to G_5 with polydispersity indices between 1.00 and 1.02, and also show massifs besides these peaks that correspond to dendrimer aggregation, whose importance increases as the generation increases (a phenomenon also observed by AFM). The steady increase of the dendrimer size can be monitored up to G_9 by AFM, with a thickness of 25 nm for this latter generation. In solution, the dendrimers have globular shapes for G_2 (81 terminal tethers) and generations higher than G_2, and their diameter can be estimated from light scattering and DOSY NMR experiments. It is necessary that, for G_6 (theoretical number of terminal tethers: $3^{6+2} = 6561$) and higher generations, the olefin termini fold back towards the center in order to avoid the peripheral bulk and fill the internal »cavities«, but backfolding is probably also already somewhat involved for generations lower than 6 [13].

References

1 (a) D.A. Tomalia, H. Baker, J. Dewald, M. Hall, G. Kallos, S. Martin, J. Roeck, J. Ryder, P. Smith, *Polym. J.* **1985**, *17*, 117; (b) D.A. Tomalia, A.M. Naylor, W.A. Goddard III, *Angew. Chem.* **1990**, *102*, 119; *Angew. Chem. Int. Ed.* **1990**, *29*, 138.

2 (a) G.R. Newkome, Z. Yao, G.R. Baker, V.K. Gupta, *J. Org. Chem.* **1985**, *50*, 2003; (b) G.R. Newkome, C.N. Moorefield, G.R. Baker, *Aldrichim. Acta* **1992**, *25*, 31; (c) G. R. Newkome, *Pure Appl. Chem.* **1998**, *70*, 2337.

3 R.G. Denkewalter, J.F. Kolc, W.J. Lukasavage, *US Patent* **1983**, 4 410 688; *Chem. Abstr.* **1984**, *100*, 103 907p .

4 (a) E. Buhleier, W. Wehner, F. Vögtle, *Synthesis* **1978**, 155; (b) F. Vögtle, G. Richardt, N. Werner, *Dendritische Moleküle–Konzepte, Synthesen, Eigenschaften, Anwendungen*, B. G. Teubner-Verlag, Stuttgart, **2007**.

5 (a) C. Hawker, J.M.J. Fréchet, *J. Chem. Soc., Chem. Commun.* **1990**, 1010; (b) C.J. Hawker, J.M.J. Fréchet, *J. Am. Chem. Soc.* **1990**, *112*, 7638.

6 T.M. Miller, T.X. Neeman, *Chem. Mater.* **1990**, *2*, 346.

7 (a) D. Astruc, J.-R. Hamon, G. Althoff, E. Román, P. Batail, P. Michaud, J.-P. Mariot, F. Varret, D. Cozak, *J. Am. Chem. Soc.* **1979**, *101*, 5445; (b) D. Astruc, *Acc. Chem. Res.* **1986**, *19*, 377; (c) N. Ardoin, D. Astruc, *Bull. Soc. Chim. Fr.* **1995**, *132*, 875.

8 D. Astruc, J.-R. Hamon, E. Román, P. Michaud, *J. Am. Chem. Soc.* **1981**, *103*, 7502.

9 H.A. Trujillo, C.M. Casado, J. Ruiz, D. Astruc, *J. Am. Chem. Soc.* **1999**, *121*, 5674.

10 D. Astruc, *Top. Curr. Chem.* **1991**, *160*, 47.

11 F. Moulines, L. Djakovitch, B. Gloaguen, W. Thiel, J.-L. Fillaut, M.-H. Delville, D. Astruc, R. Boese, *Angew. Chem.* **1993**, *105*, 1132; *Angew. Chem. Int. Ed. Engl.* **1993**, *32*, 1075.

12 V. Sartor, L. Djakovitch, J.-L. Fillaut, F. Moulines, F. Neveu, V. Marvaud, J. Guittard, J.-C. Blais, D. Astruc, *J. Am. Chem. Soc.* **1999**, *121*, 2929.

13 J. Ruiz, G. Lafuente, S. Marcen, C. Ornelas, S. Lazare, E. Clouet, J.-C. Blais, D. Astruc *J. Am. Chem. Soc.* **2003**, *125*, 7250.

14 C. Ornelas, J.R. Aranzaes, E. Cloutet, D. Astruc, *Org. Lett.* **2006**, *8*, 2751.

15 D. Astruc, *Organometallic Chemistry and Catalysis*, Springer, Berlin, **2007**, Chap. 21.

41

Three Are Better Than One:
Gathering Three Different Metals in the Same Molecule:
a Straightforward Route to 3d-3d'-4f Heterometal Systems [1]

Ruxandra Gheorghe, Augustin M. Madalan and Marius Andruh

The search for smart molecular materials is stimulating the design of a large diversity of polynuclear complexes with useful properties. The fine tuning of the physical and chemical properties can be achieved through the modular construction of a specific metal–organic system based on different metal ions which exert a structural role (directing and sustaining the solid-state architecture) and a functional one (carrying magnetic, optical, or redox properties).

Nowadays, miniaturization is increasingly demanded, and materials combining two or more desirable properties are an effective space-saving measure. Such properties arise from the presence and the interaction of the different metal ions. A strong impact of multimetallic molecular compounds on the IT industry is expected: polynuclear complexes can act as nanometric magnetic memory units for high-density information storage.

Most of the heteropolynuclear complexes contain two different metal ions, while the number of those constructed from three different metal ions is limited to only few examples. Here, we present a straightforward synthetic route towards heterotrimetallics which illustrates the bottom-up paradigm in obtaining new materials.

Apparatus	50 mL beaker, 4 × 25 mL beakers, 10 mL beaker, 5 Pasteur pipettes, weighing boat, H-shaped glass tube, cylinders, magnetic stirrers, 3 funnels, filter paper, protective gloves, safety glasses.
Chemicals	3-Methoxysalicylaldehyde, 1,3-diaminopropane, lithium hydroxide monohydrate, copper(II) acetate monohydrate, gadolinium(III) nitrate hexahydrate, potassium hexacyanoferrate(III), ethanol, acetone, distilled water.
Attention!	Safety glasses and protective gloves must be worn at all times.

Caution!   All reactions should be carried out in a well-ventilated hood.

Procedure   *Synthesis of Cu(valpn)* (where valpn^{2-} is the N,N-propylenedi(3-methoxysalicylideneiminato) anion): an aqueous solution (10 mL) of copper(II) acetate monohydrate (0.5 g, 2.5 mmol) is added with stirring to a lukewarm ethanol solution (30 mL) of 3-methoxysalicylaldehyde (0.76 g, 5 mmol). It is then deprotonated with LiOH·H$_2$O (0.21 g, 5 mmol, dissolved in 5 mL of water). 1,3-Diaminopropane (0.21 mL, 2.5 mmol) is added drop by drop to the resulting mixture. A green precipitate of Cu(valpn) appears, which is filtered off after 15 min of stirring.

Synthesis of Cu(valpn)Gd(NO$_3$)$_3$: Addition of Gd(NO$_3$)$_3$·6H$_2$O (0.9 g, 2 mmol) to Cu(valpn) (0.81 g, 2 mmol) in acetone (10 mL) induces the formation of a light green precipitate of Cu(valpn)Gd(NO$_3$)$_3$, which is filtered off.

Synthesis of [Cu(OH$_2$)(valpn)Gd(OH$_2$)$_3${Fe(CN)$_6$}]·3H$_2$O: To a solution containing Cu(valpn)Gd(NO$_3$)$_3$ (0.074 g, 0.1 mmol) in water (10 mL) is added an aqueous solution (10 mL) of K$_3$[Fe(CN)$_6$] (0.033 g, 0.1 mmol). The resulting green micro-crystalline precipitate is filtered off and washed with water. Single crystals are obtained by slow diffusion, in an H-shaped tube, of two aqueous solutions, one of them containing Cu(valpn)Gd(NO$_3$)$_3$ and the other K$_3$[Fe(CN)$_6$]. After one week, dark green crystals result, which are filtered off and washed with water.

Characterization:

Cu(valpn): IR (KBr, cm^{-1}): $\bar{\nu}$ 3481 br, 3054 vw, 2996 vw, 2900 vw, 2836 vw, 1613 vs, 1541 m, 1464 m, 1440 s, 1342 m, 1244 m, 1214 vs, 1173 w, 1102 vw, 1073 m, 965 m, 858 w, 784 vw, 739 s, 649 w, 593 vw (the most important feature being the imine C=N stretching band at 1613 cm^{-1}).

Cu(valpn)Gd(O$_2$NO)$_3$: IR (KBr, cm^{-1}): A significant difference between its IR spectrum and that of Cu(valpn) is the addition of nitrato bands at 1473 (overlapped with C=C stretching band), 1382 and 1296.

[Cu(OH$_2$)(valpn)Gd(OH$_2$)$_3${Fe(CN)$_6$}]·3H$_2$O: IR (KBr, cm^{-1}): The peaks due to the nitrato groups disappear and the presence of the cyano groups (bridging and terminal) is indicated by two new bands located at 2153 and 2130. Magnetic moment (300 K): 8.43 BM.

Waste Disposal Waste solutions containing cyanide are treated in basic solution containing hypochlorite.

Explanation

The compound $[Cu(OH_2)(valpn)Gd(OH_2)_3\{Fe(CN)_6\}] \cdot 3H_2O$, whose synthesis is described above, represents one of the results of the perpetual quest of chemists working in the area of molecular magnetism, where the interest grows with the number of metal ions and the exchange pathways between them. Although remarkable advances were made in the development of synthetic procedures, the construction of systems containing three different spin carriers remained a challenge for years because of the random mixing of metal ions in solution.

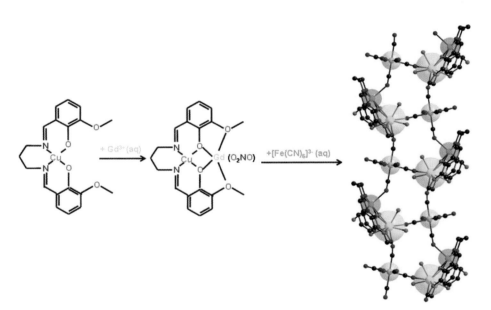

Scheme 1 Stepwise synthesis of a Cu(II)-Gd(III)-Fe(III) heterotrimetallic system (color code: gadolinium, yellow; copper, green; iron, orange; nitrogen, blue; oxygen, red; carbon, dark blue).

In order to prepare a 3d-3d′-4f heterometal system, a stepwise procedure is employed based on the self-assembly of stable heterobinuclear [Cu(II)Gd(III)] cationic species and an anionic mononuclear complex, a »metalloligand«, possessing bridging ligands able to coordinate to the copper(II) ion and/or to the 4f ion.

The synthesis of the Cu(II)-Gd(III)-Fe(III) heterotrimetallic system can be rationalized in three steps (Scheme 1). The first step is the template reaction between copper(II) acetate, 3-methoxysalicylaldehyde and 1,3-diaminopropane in a $1:2:1$ molar ratio, which yields the compartmental complex ligand Cu(valpn), where the metal ion is located into the N_2O_2 compartment. This complex possesses an outer O_2O_2 binding site that is able to accommodate the oxophilic Gd(III) ions, in the second step of the procedure, through the reaction of the Cu(valpn) complex with gadolinium(III) nitrate. The third and final step is the substitution of the nitrato groups coordinated at the Gd(III) ion with hexacyanoferrate(III) ions.

Reference

1 R. Gheorghe, M. Andruh, J.-P. Costes,
  B. Donnadieu, *Chem. Commun.* **2003**, 2778.

42

Photochemistry for Mild Metal-free Arylation Reactions

Stefano Protti, Daniele Dondi, Maurizio Fagnoni and Angelo Albini

Carbon–carbon bond formation plays a fundamental role in both biological and industrial processes. In particular, the formation of aryl–carbon bonds is of great importance [1] in the synthesis of compounds for applications ranging from pharmaceutics to optoelectronic materials. Apart from long-known processes such as the Friedel–Crafts reaction and aromatic nucleophilic substitution, which have some limitations, the development of arylations in the last two decades has mainly involved metal-mediated cross-coupling reactions [2]. As shown in Scheme 1, path *a*, these reactions could be envisaged as transition metal-assisted nucleophilic substitutions where the metal catalyst has the role of activating by oxidative addition the aryl–X or aryl–O bond (the electrophilic component of the reaction) in aryl halides or esters, respectively. Although such cross-coupling reactions are presently the most common choice, at least on a laboratory scale, their optimization in terms of environmental sustainability is somewhat limited because of the required use of toxic, aggressive and often unstable organometallic

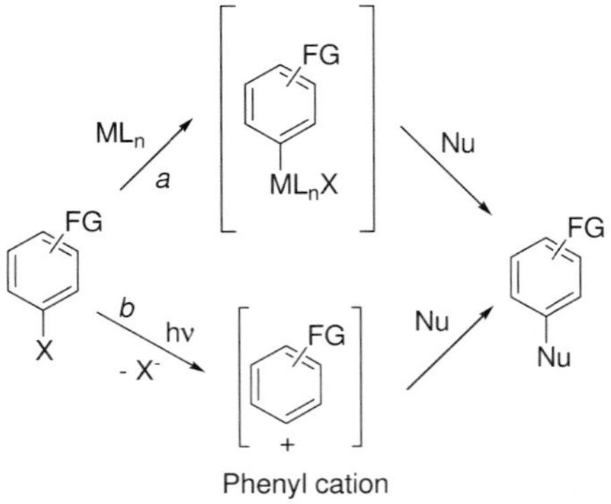

Scheme 1 Transition metal-mediated (a) and photochemical (b) arylation reactions via activation of the aryl–X bond.

derivatives as nucleophiles (Nu). In addition, elaborate workup procedures are required for the recovery of the expensive and toxic catalyst.

An appealing metal-free alternative for forming an aryl-carbon bond based on the generation of phenyl cations is shown in Scheme 1, path *b*. Thus, in polar solvents a »clean« reagent such as the photon induces the heterolytic cleavage of an aryl–halogen (or aryl–oxygen) bond in electron-rich aryl chlorides (or, respectively, esters, Scheme 1) [3]. Phenyl cations are strong electrophilic intermediates that have only recently attracted some attention in organic synthesis for the chemoselective reaction with π nucleophiles such as olefins, alkynes or (hetero)aromatics, with no interference by nucleophiles [4]. Phenyl cation chemistry has been effectively exploited for obtaining synthetically significant products belonging to the class of arylalkynes [5a], allylphenols [5b], aryl nitriles [5c] or biphenyls [6] (Scheme 2). Interestingly, the reaction can be carried out in partially aqueous media such as water/acetonitrile mixtures [5b, c].

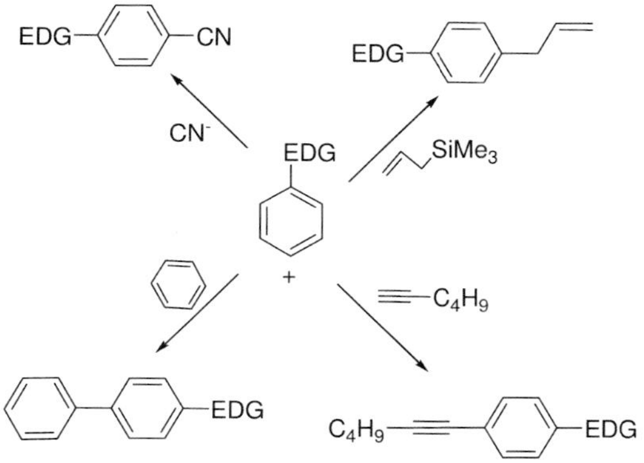

Scheme 2 Formation of aryl–C bonds via phenyl cation chemistry.
EDG = Electron-donating group.

Two representative examples of the formation of aryl–carbon bonds through photoactivation of chlorophenols under green conditions are reported below. In the first case, a sterically crowded biphenyl (**3**) was smoothly accessed by irradiation of a solution of 4-chlorophenol (**1**) in a 5:1 mixture of MeCN and water in the presence of an excess of mesitylene (**2**). A Wheland-type cation is envisaged as the intermediate (Scheme 3). Thus, a direct clean *mono*-arylation of aromatics is attained with no need of a metal catalyst. Furthermore, the process is strictly chemoselective, and no functionalization of benzylic hydrogens took place.

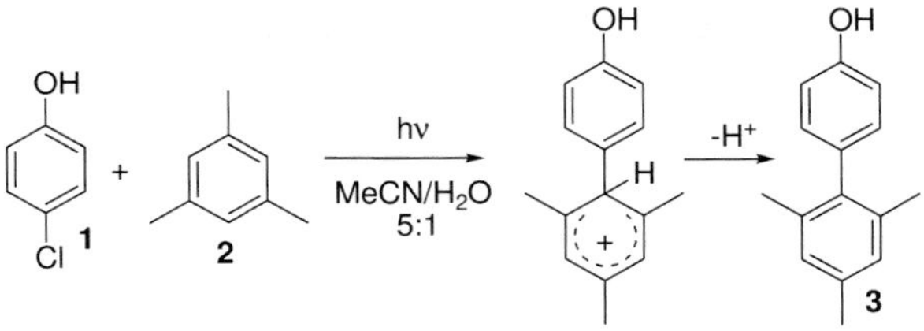

Scheme 3 Synthesis of biphenyl **3** by photochemically direct arylation of mesitylene (**2**).

Benzyl and phenyl tetrahydrofurans represent another synthetic target that is easily achieved by these photoinduced arylations [7]. Addition of a 2-hydroxy-phenyl cation (from **4**) onto the double bond of an alkenol (e.g. 4-penten-1-ol, **5**) generates an adduct carbocation and the benzyl-tetrahydrofuran **6** from it via intramolecular nucleophilic addition of the hydroxy group of the alcohol in a one-pot reaction (Scheme 4). It is noteworthy that, after the first C–C bond formation, a C–O bond formation with the alcoholic OH occurs exclusively, with no competitive addition of solvent water or intramolecular addition by the phenol function.

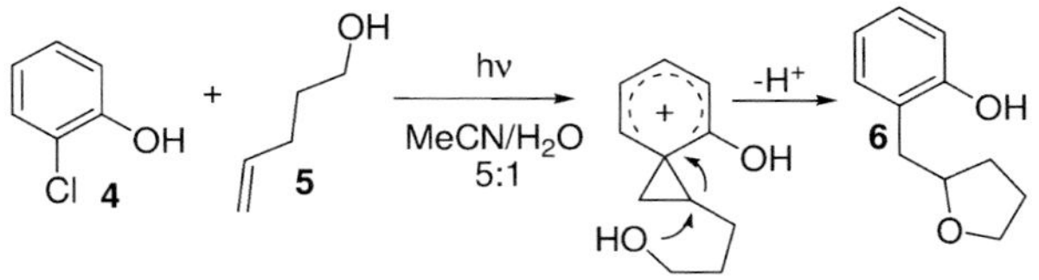

Scheme 4 Synthesis of 2-benzyl-tetrahydrofuran (**6**) by photochemical reaction between 2-chlorophenol (**4**) and 4-penten-1-ol (**5**).

To summarize, the chemistry of phenyl cations is complementary to metal-catalyzed cross-coupling reactions and has some advantage from the environmental point of view. In fact, this is effective with easily available electron-rich aryl chlorides (usually less reactive under palladium catalysis) and does not require strictly anhydrous conditions or a high temperature. In addition, no expensive catalysts and ligands are used, and the reaction involves simple alkenes or aromatics as nucleophiles in the place of the more aggressive and/or less easily available organometallic derivatives. As demonstrated by the chlorophenol reaction above, the mild conditions of the photochemical process enable one to skip the protection/deprotection steps of the OH group that would be required under thermal conditions involving strong bases or nucleophiles.

Laboratory experiments to demonstrate the reactions described above

Synthesis of 4-hydroxy-2′,4′,6′-trimethylbiphenyl (3)

Apparatus	Nitrogen-purged solutions in quartz tubes were irradiated in a multilamp reactor fitted with six 15-W phosphor-coated lamps (maximum of emission at $\lambda = 310$ nm).
Chemicals	4-chlorophenol (1) and mesitylene (2) were purchased from Sigma-Aldrich and freshly purified before use. Acetonitrile and water used were of HPLC grade purity. Cyclohexane and ethyl acetate used in the purification step were freshly distilled before use.
Attention!	Safety UV-protective glasses, a laboratory coat and protective gloves must be worn at all times during this experiment.
Caution!	Both the preparation of the solutions and the purification of the raw product should be carried out in a well-ventilated fume hood.
Procedure	A solution of 4-chlorophenol (1, 770 mg, 6 mmol) and mesitylene (2, 3.00 mL, 30 mmol) in a water/acetonitrile 1:5 mixture (60 mL) was degassed by nitrogen bubbling, then irradiated at 310 nm for 16 h. The photolyzed solution was evaporated and the resulting residue purified by column chromatography (eluent: cyclohexane/ethyl acetate, 9:1), affording 751 mg of 4-hydroxy-2′,4′,6′-trimethylbiphenyl (3), white solid, 59% yield, mp: 142–143 °C, lit.[8] 148 °C).

Characterization data: ^{1}H NMR (300 MHz, CDCl$_3$): δ 8.30 (bs, 1H), 6.90-7.00 (m, 6H), 2.30 (s, 3H), 2.00 (s, 6H) ppm. ^{13}C NMR (75 MHz, CDCl$_3$): δ 156.9 (C$_q$), 139.9 (C$_q$), 136.8 (C$_q$), 133.4 (C$_q$), 131.2 (CH), 129.0 (CH), 116.4 (CH), 21.5 (CH$_3$), 21.4 (CH$_3$) ppm. IR (neat, cm^{-1}): $\bar{v}$ 3500, 2924, 1458, 1259, 836. Anal. Calcd. for C$_{15}$H$_{16}$O: C 84.87, H 7.60. Found: C 84.8, H 7.7.

Waste disposal Organic solvents, reactants and exhausted silica gel were separately collected and disposed of.

Synthesis of 2-(2-hydroxyphenyl)-methyltetrahydrofuran (6)

Apparatus Nitrogen-purged solutions in quartz tubes are irradiated by using six external 15 W phosphor-coated lamps (maximum of emission at $\lambda = 310$ nm).

Chemicals 2-Chlorophenol (4) and 4-penten-1-ol (5) were purchased from Sigma-Aldrich and freshly purified by distillation before use. Acetone, acetonitrile and water were of HPLC grade purity. Cyclohexane and ethyl acetate used in the purification step were freshly distilled before use.

Attention! Safety UV-protective glasses, a laboratory coat and protective gloves must be worn at all times during this experiment.

Caution! Both the preparation of the solutions and purification of the raw product should be carried out in a well-ventilated fume hood.

Procedure A solution of 2-chlorophenol (4, 153 µL, 1.5 mmol, 0.05 M), acetone (3.00 mL, 27 mmol, 0.9 M) and 4-penten-1-ol (5, 1.55 mL, 15 mmol) in a water/acetonitrile 1:5 mixture (30 mL) was deoxygenated by nitrogen bubbling, then irradiated at 310 nm for 24 h. The photolyzed solution was evaporated and the resulting residue purified by column chromatography (eluent: cyclohexane/ethyl acetate 9:1), affording 168 mg of 2-(2-hydroxyphenyl)-methyltetrahydrofuran (6, [9] oil, 63% yield).

Characterization data: ^{1}H NMR [10] (300 MHz, CDCl$_3$): δ 8.70 (bs, 1H), 7.10–7.20 (dt, $J = 2$ and 7 Hz, 1H), 7.00–7.05 (dd, $J = 2$ and 7 Hz, 1H), 6.90–7.00 (d, $J = 7$ Hz, 1H), 6.80–6.90 (dt, $J = 2$ and 7Hz, 1H), 4.10–4.20 (m, 1H), 3.70–3.80 (m, 1H), 3.85–3.95 (m, 1H), 3.10–3.15 (dd, $J = 3$ and 14 Hz, 1H), 2.70–2.80 (dd, $J = 6$ and 15 Hz,

1H) 1.80–2.05 (m, 3H), 1.60–1.70 (m, 1H) ppm. ^{13}C NMR (75 MHz, CDCl$_3$): δ 155.9 (C$_q$), 131.7 (CH), 128.3 (CH), 124.8 (C$_q$), 119.9 (CH), 117.3 (CH), 80.8 (CH), 68.4 (CH$_2$), 34.4 (CH$_2$), 29.8 (CH$_2$), 25.9 (CH$_2$) ppm. IR (neat, cm^{-1}): $\bar{v}$ 3270, 2875, 1489, 1245, 1045, 755. Anal. Calcd. for C$_{11}$H$_{14}$O$_2$: C 74.13, H 7.92. Found: C 74.1, H 7.8.

Waste disposal Organic solvents, reactants and exhausted silica gel were separately collected and disposed of.

References

1 L. Ackermann, *Modern Arylation Reactions*, Wiley-VCH, Weinheim, **2009**.
2 A. de Meijere, F. Diederich, *Metal-Catalyzed Cross-Coupling Reactions*, Wiley-VCH, Weinheim, **2004**.
3 M. Fagnoni, A. Albini, *Acc. Chem. Res.* **2005**, *38*, 713.
4 V. Dichiarante, M. Fagnoni, *Synlett* **2008**, 787.
5 (a) S. Protti, M. Fagnoni, A. Albini, *Angew. Chem. Int. Ed.* **2005**, *44*, 5675; *Angew. Chem.* **2005**, *117*, 5821; (b) S. Protti, M. Fagnoni, A. Albini, Org. Biomol. Chem. **2005**, 2868; (c) V. Dichiarante, M. Fagnoni, A. Albini, *Chem. Commun.* **2006**, 3001.
6 V. Dichiarante, M. Fagnoni; A. Albini, *Angew. Chem. Int. Ed.* **2007**, *46*, 6495, *Angew. Chem.* **2007**, *119*, 6615.
7 S. Protti, D. Dondi, M. Fagnoni, A. Albini, *Eur. J. Org. Chem.* **2008**, 2240.
8 G. Häfelinger, F. Hack, G. Westermayer, *Chem. Ber.* **1978**, *111*, 1323.
9 P. Normant, *Bull. Chim. Soc. Fr.* **1938**, *5*, 1148.
10 H.J. Roth, K. Michel, *Arch. Pharm. Ber. Dtsch. Pharm. Ges.* **1971**, *304*, 278.

43

An Experiment to Demonstrate the Greenhouse Effect [1]

Herbert W. Roesky

The amount of carbon dioxide generated annually by burning fossil fuels and calcining limestone for the production of cement is increasing steadily (see Figure 1). In spite of all the efforts made by discerning scientists, engineers and businessmen it is highly unlikely that this emission will cease in the near future. However, model calculations predict that if the concentration of CO_2 in the atmosphere rises to double its present value there will be an increase of between 1.5 and 4.5 °C in the temperature of the atmosphere at the earth's surface; this is likely to have a dramatic influence on vegetation, climate and the level of the oceans.

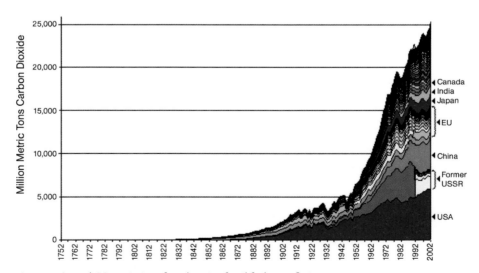

Figure 1 Annual CO_2 emissions from burning fossil fuels, gas flaring, and calcining limestone for the manufacture of cement (from: Carbon Dioxide Information Analysis Center).

Apparatus	Two 250 mL beakers, two brass discs 0.5 mm thick and 63 mm in diameter, lamp with shade and 100 W bulb, coaxial Ni/CrNi thermocouple connected to both beakers, thermometer, glass plate, gas inlet tube, safety glasses, protective gloves.
Chemicals	CO_2 from a pressure cylinder, 250 mL of a 10% NaOH solution which contains about 2.5 g $K_2S_2O_8$.
Procedure	The two brass discs are blackened for 2 h in the potassium peroxodisulfate solution. They are then laid at the bottom of the two beakers, which are standing next to one another on the demonstration table. The lamp is switched on, and the two beakers are illuminated for a few minutes. The temperature difference between the beakers is 0 ± 2 °C. One of the beakers is slowly filled with the greenhouse gas CO_2. The second beaker is covered with the glass plate while the gas is being passed in. After about 30 s the gas inlet tube and the glass plate are removed. The lamp is now switched on again and the two beakers illuminated. Shortly afterwards a temperature difference can be clearly measured: the beaker filled with CO_2 increases in temperature by about 10 °C within a minute. The temperature difference then falls because of convection and diffusion of the carbon dioxide. The blackened brass discs, which re-emit the visible light absorbed in the form of infrared radiation, are vital in this experiment; no temperature difference is generated if they are absent. The determination of the temperature change can be carried out either with a thermocouple or with a normal thermometer. However, in the latter case it is hard to measure the temperatures and no longer possible to compare them directly. The carbon dioxide concentration in the atmosphere is of course very much lower than in our experiment. The CO_2 could be replaced by CF_2Cl_2 or other chlorofluorocarbons which have a dipole moment. The experimental set-up is shown diagrammatically in Figure 2.
Waste Disposal	The alkaline solution of the oxidizing agent can be poured down the drain after being well diluted.

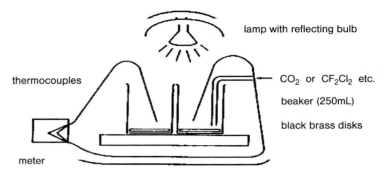

lamp with reflecting bulb

thermocouples

CO_2 or CF_2Cl_2 etc.

beaker (250mL)

black brass disks

meter

Figure 2 Experimental set-up for the simulation of the greenhouse effect.

Explanation

Carbon dioxide is at a globally averaged concentration of approximately 387 ppm in the earth's atmosphere, although in our experiment we use pure carbon dioxide to magnify its effect. Carbon dioxide transmits visible light. Therefore the visible light from the lamp is not absorbed by the carbon dioxide. However, the black metal plates absorb the visible light and emit infrared and near-infrared light. This light is then strongly absorbed by the carbon dioxide, causing its temperature to increase. Therefore carbon dioxide is called a greenhouse gas.

Reference

1 M. Adelhelm. E.-G. Höhn, D.A. Franz,
   J. Chem. Educ. **1993**, *70*, 73.

44

Organic Synthesis through Semiconductor Photocatalysis

Horst Kisch

In the recent work on the use of metal sulfide powders as heterogeneous photocatalysts for the synthesis of new organic compounds, two general types of reaction pathways become apparent. Whereas all reactions are initiated by primary electron transfer between the excited semiconductor surface and adsorbed substrates, the further transformation to the final products falls into two categories. In the most frequently observed *Type A* reactions *two* final redox products are obtained, analogous to photoelectrolysis and conventional electrolysis. A typical example is the formation of hydrogen and dehydrodimers upon irradiation of zinc sulfide in the presence of olefins or enol/allyl ethers. In the very rare case of *Type B* reactions only *one* final product is formed, as observed in the cadmium sulfide-catalyzed photoaddition of 1,2-diazenes or imines to olefins or enol/allyl ethers. This transformation has no counterpart in photoelectrochemistry and can be viewed as *paired photoelectrolysis*. This simple preparation of a pharmaceutically important class of organic compounds demonstrates the usefulness of semiconductor photocatalysis in chemical syntheses.

Addition of α-pinene to *N*-phenylbenzophenone imine

Apparatus   Solidex immersion lamp apparatus (220 mL) equipped with a tungsten halogen lamp (Osram 100 W, 12 V, $\lambda > 350$ nm), safety sun glasses, laboratory coat, protective gloves.

Chemicals   Cadmium sulfide prepared under dinitrogen, *N*-phenylbenzophenone imine, α-pinene.

Procedure   *N*-phenylbenzophenone imine (1.55 g, 5.82 mmol), CdS [1] (0.30 g, 2.08 mmol), and α-pinene (0.23 mol, 37.0 mL) are suspended in

MeOH(200 mL) in the immersion lamp apparatus and sonicated for 20 min under Ar bubbling. Subsequent irradiation with a tungsten halogen lamp is stopped when all of the imine is consumed, as indicated by thin-layer chromatography. Typical reaction times are 20–22 h. CdS is removed by suction filtration, and the remaining liquid is evaporated *in vacuo*. The resulting white powder is recrystallized from *n*-heptane. Yield: 1.64 g (72%) white solid.

Characterization data: m.p.156–159 °C. ^{1}H NMR (CDCl$_3$, TMS, 270 MHz) δ = 0.15 (d, 1H, =C(C)–CH(C)–HCH–), 0.90 (s, 3H, CH_3–(C)C(C)–CH$_3$), 1.20 (s, 3H, CH$_3$–(C)C(C)–CH_3), 1.60 (s, 3H, CH_3–C=CH–), 1.90 (m, br, 2H, =C(C)–CH(C)–HCH–), 3.35 (s, br, 1H, –CH–C–N–), 4.50 (s, 1H, NH), 5.10 (s, 1H, –C(C)=CH–CH (C)–), 6.30 – 6.90 (m, 5H, –N–C$_6$H$_5$), 7.10 – 7.55 (m, 10H, –C–C$_6$H$_5$) ppm. ^{13}C NMR (CDCl$_3$, 67.7 MHz) δ = 20.63 (CH_3–(C)C(C)–CH$_3$), 23.27 (CH_3–C=CH–), 25.53 (=C(C)–CH(C)–CH_2–), 26.52 (CH_3–(C)C(C)–CH$_3$), 42.34 (=C(C)–CH(C)–CH_2–), 42.48 (=C(C)-CH(C)-(C)C(C)-), 46.60 (=CH-CH(C)-CH(C)-), 52.15 (=CH–CH (C)–), 68.23 (-C-N-C$_6$H$_5$), 115.74 (–C(C)=CH-), 116.36, 117.33, 128.23 (–N–C_6H$_5$), 126.58, 127.36, 127.63, 128.82, 129.70 (–C–C_6H$_5$), 140.72, 142.58 (–C–C–N, C_6H$_5$), 146.01 (–C–N, C$_6$H$_5$), 147.44 (–C(C)=CH–) ppm. IR (KBr) [cm^{-1}]: 3387 (NH), 3033, 3002, 2976, 2949 (CH), 1598, 1502 (C=C); MS: 393 [M$^+$]; C$_{29}$H$_{31}$N (393.57); calcd. C: 88.50, H: 7.94, N: 3.56; found C: 88.15, H: 8.05, N 3.56.

Explanation

In the proposed mechanism of the addition reaction [Eq. (1)] visible light absorption by

$$(1)$$

cadmium sulfide generates an electron-hole pair which is trapped and separated into the reducing and oxidizing surface centers e_r^- and h_r^+. Proton-coupled reduction and oxidation

$$Ar_2C = N - Ar + H_3O^+ + e_r^- \rightarrow Ar_2C - N(H)Ar + H_2O \tag{2}$$

$$RH + h_r^+ + H_2O \rightarrow R^· + H_3O^+ \tag{3}$$

$$Ar_2C - N(H)Ar + R^· \rightarrow Ar_2C(R) - N(H)Ar \tag{4}$$

afford an α-aminobenzyl and allyl (R) radical, respectively [Eqs. (2, 3)]. Hetero-coupling of the two radicals leads to the final product (Eq. 4). Thus, the most significant reaction steps of this novel type of photoaddition reaction are a primary electron transfer followed by stereoselective radical C–C coupling.

Reference

1 H. Keck, W. Schindler, F. Knoch, H. Kisch, *Chem. Eur. J.* **1997**, *3*, 1638 and references cited therein.

45

Daylight Semiconductor Photocatalysis for Air and Water Purification

Horst Kisch

Semiconductor photocatalysis is an efficient method for the chemical utilization of solar energy. It is based on the surface trapping of light-generated charges, which induce interfacial electron transfer reactions with a great variety of substrates. Titania, the most promising photocatalyst, is already used in practical applications like window panes and self-cleaning paints. However, because of its large bandgap of 3.20 eV, only the small fraction of solar UV light (about 2–3%) can be utilized. Therefore, many attempts have been made to shift the photocatalytic activity of titania to the much larger visible spectral region. In previous work we have shown that incorporation of a few weight percent of a transition metal chloride either in the bulk or at the surface of titania leads to visible light photocatalysis, enabling complete oxidation (photomineralization) of the ubiquitous pollutant 4-chlorophenol to carbon dioxide and hydrochloric acid by visible light ($\lambda \geq 455$ nm). We recently found that a coke containing titania powder was active with visible as well as UV light. This material was prepared by a sol-gel method using various titanium alkoxide precursors. Since the photocatalytic activity, especially at low light intensity, was much smaller than that of platinum(IV) chloride-modified titania, we searched for improvement. This was achieved by hydrolyzing titanium tetrachloride or other titania precursors with a carbon-containing base like tetrabutylammonium hydroxide or by simple calcination of titania with a carbon-containing material. The latter method is already applied in industry, and the resulting TiO_2/C photocatalyst induces removal of pollutants from air and water through photooxidation by aerial oxygen.

Pollutant oxidation in air and water

Apparatus
100 mL Erlenmeyer flask, magnetic stirrer, 1 L round bottomed flask, filter paper (11 cm diameter), safety glasses, protective gloves, pH meter, IR instrument.

Chemicals
4-Chlorophenol, acetaldehyde, benzene, carbon monoxide. TiO_2/C: commercial product VLP 7000 from *Kronos* or prepared according to Ref. [1].

Attention!
4-Chlorophenol, acetaldehyde, benzene and carbon monoxide are very toxic and should be handled in a well-ventilated hood.

Procedure
Water purification: In a 100 mL Erlenmeyer flask, TiO_2/C (50 mg) is suspended in aqueous 4-chlorophenol solution (50 mL, 2.5×10^{-4} mol L^{-1}). The flask is placed on a bench close to a window, allowing the absorption of diffuse daylight (4–6 W m^{-2} at 400–1200 nm). Mineralization of 4-chlorophenol is followed by measuring the total organic content or decrease in the pH-value.

Air Purification: An aqueous suspension of TiO_2/C (12 mg) is applied to the filter paper followed by drying at 80 °C. The photocatalyst-loaded paper so obtained is inserted into the round-bottomed flask. Acetaldehyde or benzene in vapor form or carbon monoxide gas are added with a syringe at an appropriate concentration, resulting in a final pollutant content of 5 vol%. Carbon dioxide formation is monitored by IR gas phase spectroscopy.

Explanation

Because of its diradical nature, molecular oxygen has to be activated by energy or electron transfer to enable it to perform fast oxidation reactions. The resulting singlet oxygen and superoxide, respectively, are known as reactive oxygen species (ROS), and these can bring about the complete oxidation of many organic and inorganic compounds. In most cases, activation is induced photochemically. Reactions of this type may be responsible for the »self cleaning« of natural waters. In the present case, the reaction [eq. (1)] is started by the daylight-induced generation of an electron-hole pair, which is trapped and separated into the reducing and oxidizing surface centers e_r^- and h_r^+.

$$2\,p\text{-ClC}_6\text{H}_4\text{OH} + 13\,\text{O}_2 \rightarrow 12\,\text{CO}_2 + 2\,\text{HCl} + 4\,\text{H}_2\text{O} \tag{1}$$

Electron transfer from the semiconductor to oxygen produces superoxide and eventually ROS [eq. (2)], whereas dissociative electron transfer from 4-chlorophenol to the semiconductor affords an aryl radical [eq. (3)].

$$\text{O}_2 + \text{e}_r^- \rightarrow \text{O}_2^{\cdot -} \rightarrow \ \rightarrow \text{ROS} \tag{2}$$

$$\text{ArOH} + \text{h}_r^+ + \text{H}_2\text{O} \rightarrow \text{Ar}^\cdot + \text{H}_3\text{O}^+ \tag{3}$$

$$\text{Ar} + \text{ROS} \rightarrow \text{CO}_2 + \text{HCl} \tag{4}$$

A subsequent multi-step reaction cascade finally leads to complete mineralization [eq. (4)]. The mechanism of the gas phase oxidations can be formulated analogously. In summary, absorption of daylight by the semiconductor generates reactive charges capable of activating molecular oxygen and pollutant through interfacial photoredox catalysis.

Reference

1 (a) S. Sakthivel, H. Kisch, *Angew. Chem. Int. Ed.* **2003**, *42*, 4908; *Angew. Chem.* **2003**, *115*, 5057; (b) H. Kisch, P. Zabek, J. Eberl, *Photochem. Photobiol. Sc.*, in press.

46

An Evergreen in CO_2-Fixation: the Kolbe–Schmitt Reaction

Sandra Walendy, Giancarlo Franciò, Markus Hölscher and Walter Leitner

Carbon dioxide (CO_2) is the most abundant C1 building block available for chemical syntheses. The earth's atmosphere contains ca. 762 Gt of carbon, most of which is present as carbon dioxide. Each year natural processes lead to an exchange of ca. 210 Gt of carbon between the atmosphere, the biosphere and the hydrosphere. Man-made CO_2 emissions, primarily generated by the burning of fossil fuels, is currently in the 30 Gt range. This »waste material« from human activities is often available in very concentrated form as flue gases. For example, emissions from power stations or chemical processes such as ethylene oxide production. Although the use of this waste stream in a potential chemical process would only have a marginal direct impact on anthropogenic CO_2 emissions, such a use is an attractive option within the framework of a general carbon management strategy. Currently, CO_2 is already used in large scale industrial production in the synthesis of urea from CO_2 and NH_3 (ca. 80 Mt a^{-1}) and the synthesis of methanol from syngas (ca. 2 Mt a^{-1}). There is an estimated potential for a further reduction by ca. 2 Gt a^{-1} of atmospheric CO_2 by introducing novel synthesis routes and products, many of which involve transition metal-catalyzed reactions [1]. A large variety of interesting transformations including pathways to formic acid, its derivatives [2], monomeric and polymeric organic carbonates, and many other C–C– and C–O– bond-forming reactions have been successfully demonstrated on a laboratory scale [3].

A classic and very illustrative example of the introduction of the CO_2 molecule into a chemical compound is the Kolbe–Schmitt synthesis [4], which was originally developed in 1860 by Kolbe and Lautemann during the synthesis of salicylic acid (2-hydroxybenzoic acid). Phenol and sodium metal were heated in an atmosphere of carbon dioxide, generating the desired product in varying yields. In 1884 Schmitt exposed dry sodium phenoxide to a high CO_2 pressure in a sealed tube and heated it up to 100 °C [5]. This procedure gave a quantitative conversion and increased the yields of the salicylic acid derivatives significantly.

The reaction was applied to various substrates in later years. Today the commercial production of salicylic acid reflects the importance of the Kolbe–Schmitt reaction in industry. The reliable literature protocol described here for the carboxylation of thymol [6, 7] for the synthesis of *o*-carvacrotinic acid represents an excellent example for introductory laboratory courses in undergraduate studies.

Preparation of 2-hydroxy-6-isopropyl-3-methylbenzoic acid (*o*-carvacrotinic acid)

Apparatus	The following procedure was carried out under a CO_2 atmosphere in a three-necked round-bottomed flask (500 mL) equipped with a magnetic stirring bar, a gas inlet tube ending with a frit, and a high-efficiency coiled and jacketed condenser with an oil-filled bubbler at the top.

Heating and stirring plate with a thermo-element, oil bath, CO_2 cylinder with precision adjusting valve, connecting tube, argon cylinder with reducing valve, vacuum pump, safety glasses, laboratory coat, and protective gloves.

Chemicals   2-Isopropyl-5-methylphenol (trivial name: thymol), sodium metal, CO_2, *o*-xylene, molecular sieves, water, *i*-propanol, aqueous HCl (32%), aqueous $NaHCO_3$ (5%), *n*-hexane.

Attention!   Safety glasses, a laboratory coat and protective gloves must be used at all times.

Caution!   The reaction should be carried out in a well-ventilated fume hood. Sodium reacts violently with water.

Procedure   2-Isopropyl-5-methylphenol (**1**) (30 g, 0.2 mol) was introduced as a solid into a three-necked flask under an argon atmosphere. Degassed and dried (drying agent: molecular sieves; residual water content: 2.5 ppm) o-xylene (375 mL) was added, resulting in a colorless solution. Sodium metal (10.0 g, 0.43 mmol) was added in small portions over 5 h. The resulting red-brown suspension was heated at 120 °C and purged with a vigorous stream of CO_2 for 20 h. The mixture was cooled to ambient temperature and the CO_2 lance was disconnected. Excess sodium was destroyed by careful addition of 2-propanol (50 mL) followed by water (350 mL). The organic layer was extracted four times with 25 mL portions of $NaHCO_3$ (5%). The combined aqueous extracts were acidified to pH 1 with concentrated HCl resulting in the formation of a brown precipitate. The precipitate was collected and dried. Recrystallization from *n*-hexane (90 mL) gave **2** (23.37 g, 74%) as reddish solid.

Characterization data: ^1H NMR (300 MHz, $CDCl_3$): δ 11.41 (s, 1H), 8.48 (s, br, 1H), 7.28 (d, $J = 7.7$ Hz, 1H), 6.72 (d, $J = 7.7$ Hz, 1H), 3.34 (sept, $J = 6.9$ Hz, 1H,), 2.58 (s, 3H), 1.22 (d, $J = 6.9$ Hz, 6H) ppm. $^{13}C\{^1H\}$ NMR (75 MHz, $CDCl_3$): δ 177.1, 161.3, 139.9, 134.7, 132.1, 122.7, 110.4, 26.5, 24.1, 22.3 ppm.

Waste Disposal The organic residues are collected and disposed of in the container for organic waste.

References

1 W. Leitner, *Coord. Chem. Rev.* **1996**, *153*, 257–284.
2 W. Leitner, *Angew. Chem.* **1995**, *107*, 2391–2405; *Angew. Chem. Int. Ed. Engl.* **1995**, *34*, 2207–2221.
3 T. Sakakura, J.-C.Choi, H. Yasuda, *Chem. Rev.* **2007**, *107*, 2365–2387.
4 (a) H. Kolbe, E. Lautemann, *Liebigs Ann. Chem.* **1860**, *113*, 125–127; (b) H. Kolbe, E. Lautemann, *Liebigs Ann. Chem.* **1860**, *115*, 157–206; (c) H. Kolbe, E. Lautemann, *Liebigs Ann. Chem.* **1860**, *115*, 178.

5 (a) R. Schmitt, German Patent 29939, **1884**, 233; (b) R. Schmitt, *J. Prakt. Chem./Chem.-Ztg.* **1885**, *31*, 397.
6 Experimental procedure adapted from: J. M. Gnaim, B. S. Green, R. Arad-Yellin, P. M. Keehn *J. Org. Chem.* **1991**, *56*, 4525–4529.
7 For a recent mechanistic study see: Y. Kosugi, Y. Imaoka, F. Gotoh, M. A. Rahim, Y. Matsui, K. Sakanishi *Org. Biomol. Chem.* **2003**, *1*, 817–821.

Index

Rose Bengal (4,5,6,7-tetrachloro-2′,4′,5′,7′-tetraiodofluorescein) 199ff.
[RuCl₂(p-cymene)]₂ 65ff.

s

S-arylation 14
sarcophagine (3,6,10,13,16,19-hexa-azabicyclo[6.6.6]icosane) 225
Schrock carbene complex 30
self cleaning 272f.
semiconductor 238ff.
  – daylight 272
  – photocatalysis 269
silicas
  – aminopropylated (APS) 7ff.
silicon
  – encapsulated 149f.
sodium
  – recycling 212
  – residue disposal 197
solid acid catalyst 3
  – activity 5
solvent extraction (SX) 92
SPO (secondary phosphine oxide) 65
styrene 103, 115ff.
superweak anion 131
surface passivation 241
synthetic transformation 149ff.

t

template synthesis 225
terephthalic acid 63
tetra-n-butylammonium bromide (TBAB) 104, 106
tetra-n-butylammonium perchlorate (TBAP) 80ff.
tetrachloro(N,N′-diisopropylchloroamidino)phosphorus(V)
  – preparation 147
tetraethoxysilane (TEOS) 81, 151
thiol 73f.
  – copper-catalyzed arylation 11ff.
titanacyclopropane 173

titania
  – platinum(IV) chloride modified 272
titanium catalyst 206
p-triallylmethylphenol dendron 247ff.
triazole 66, 178
trichlorohydridoaluminate species 125, 127
triethylsulfonium hexachlorobromodialuminate [Et₃S][Al₂Cl₆Br] 124
(R)-3,3,3-trifluoro-2-(indol-3-yl)-2-hydroxyl-propionic acid ethyl ester 32
(S)-3,3,3-trifluoro-2-(indol-3-yl)-2-hydroxyl-propionic acid ethyl ester 32
trimethylalane (AlMe₃) 230f., 160ff.
1,3,5-triphenylpyrazole
  – synthesis 45ff.
tris(ethylenediamine)nickel(II) tris(catecholato)silicate 156
1,3,5-tris(triallylmethyl)benzene 250

u

Uncaria rhynchophylla (Miq.) 160
urdamycinone B 199

v

vitamin E 161
volatile organic compound (VOC) 86, 92, 108

w

Wacker–Heck reaction 162f.
Wacker-type cyclization 162
water purification 272ff.
water-ammonia 73ff.
weakly coordinating anion (WCA) 131
  – synthesis 132
Wheland-type cation 261
Williamson reaction 247

z

zeolite 4f., 7, 40, 229
zirconia
  – sulfated 4f.
Zn[Au(CN)₂]₂ 87ff.
Zurich model 186